P9-CTQ-561

BF 591 .V47 2002

The verbal communication of
emotions

DATE DUE

The Verbal Communication of Emotions

Interdisciplinary Perspectives

NEW ENGLAND INSTITUTE
OF TECHNOLOGY
LIBRARY

The Verbal Communication of Emotions

Interdisciplinary Perspectives

Edited by

Susan R. Fussell
Carnegie Mellon University

NEW ENGLAND INSTITUTE OF TECHNOLOGY
LIBRARY

LAWRENCE ERLBAUM ASSOCIATES, PUBLISHERS
2002   Mahwah, New Jersey   London

3|03   # 47838360

Camera ready copy for this book was provided by the author.

Copyright © 2002 by Lawrence Erlbaum Associates, Inc.
All rights reserved. No part of this book may be repro-
duced in any form, by photostat, microform, retrieval system, or any other means, without prior written permission
of the publisher.

Lawrence Erlbaum Associates, Inc., Publishers
10 Industrial Avenue
Mahwah, NJ 07430

Cover design by Kathryn Houghtaling Lacey

Library of Congress Cataloging-in-Publication Data

The verbal communication of emotions : interdisciplinary
    perspectives / edited by Susan R. Fussell.
            p. cm.
Includes bibliographical references and index.
ISBN 0-8058-3689-6 (alk. paper)
ISBN 0-8058-3690-X (pbk. : alk. paper)
1. Expression. 2. Emotions. I. Fussell, Susan R.
BF591 .V47 2002
153.6—dc21   2002040792
                                                    CIP

Books published by Lawrence Erlbaum Associates are
printed on acid-free paper, and their bindings are chosen for
strength and durability.

Printed in the United States of America
10 9 8 7 6 5 4 3 2 1

To Frank, for all his love and encouragement throughout this project and all areas of my life.

Contents

– 1 –
The Verbal Communication of Emotion: Introduction and Overview

Susan R. Fussell
Carnegie Mellon University

The interpersonal communication of emotional states is fundamental to both everyday and clinical interaction. One's own and others' affective experiences are frequent topics of everyday conversations, and how well these emotions are expressed and understood is important to interpersonal relationships and individual well-being. Similarly, in therapeutic contexts, progress depends on, among other things, how articulately the client expresses his or her emotions and how well the therapist understands and responds to these expressions. In this volume we take an interdisciplinary approach to understanding the verbal communication of emotion in a variety of contexts.

All languages provide speakers with an array of verbal strategies for conveying emotions. In English, for example, we have an abundance of both literal (e.g., *irked, angry, furious*), and figurative (e.g., *flipping one's lid, blow a gasket*) expressions that can be used to describe a theoretically infinite number of emotional states (e.g., Bush, 1973; Clore, Ortony, & Foss, 1987; Davitz, 1969; Johnson-Laird & Oatley, 1989; Ortony, Clore, & Collins, 1988; Ortony, Clore, & Foss, 1987). Studies of language use in psychotherapy likewise are replete with examples of literal and figurative expressions for emotions (e.g., Angus, 1996; Davitz, 1969; Davitz & Mattis, 1964; Ferrara, 1994; Karp, 1996; McMullen & Conway, 1996; Pollio & Barlow, 1975; Siegelman, 1990).

This book pulls together new research and theory on the verbal communication of emotions by an international, cross-disciplinary group of recognized experts in affective communication, with the goal of providing readers with a comprehensive view of current research and fertilizing cross-disciplinary interaction. Topics include analyses of literal and figurative expressions for emotions, studies of the use of metaphor and other figurative expressions for emotion, analysis of the role of conversational partners in creating emotional meaning, and the effects of culture on emotional communication. In the remainder of this introductory chapter, I first describe the scope of the book; then, I briefly summarize the chapters in each section of the book; finally, I describe several themes and issues that arise throughout the book and outline some areas for future research.

1

THE SCOPE OF THE BOOK

The field of emotional communication is very large; comprehensive coverage of all approaches to this topic would far exceed the page limits of this book. In this section I briefly describe the scope of the volume.

Verbal Communication of Emotions

First, all contributions deal explicitly with the *verbal* communication of emotions. It is well established that humans use a wide range of nonverbal and paralinguistic mechanisms to express emotion, including facial expressions, gestures, posture, tone of voice, and the like. Over the past several decades, substantial progress has been made in understanding how emotions are expressed through these nonverbal mechanisms (see, e.g., papers in Barrett, 1998; Ekman & Davidson, 1994, Feldman & Rimé, 1991; Philippot, Feldman, & Coats, 1999; Russell & Fernandez-Dols, 1997; Scherer & Ekman, 1984).

Important as these modalities are, however, paralinguistic and nonverbal channels in and of themselves are insufficient for expressing the full range of human emotional experiences for several reasons. First, although nonverbal cues can indicate what general class of emotions a person is feeling, they typically do not provide detailed information about that person's emotional state. By seeing that someone is crying, for instance, we might assume that they are sad; by the extent of sobbing we might even be able to infer the intensity of the sadness. But the tears in and of themselves provide no information about the particular experience of sadness, for example, the cognitions that go along with the sadness (e.g., "I have no money" vs. "I'm lonely") or the circumstances that lead up to feeling sad (e.g., "I lost my job" vs. "My dog just died"). As the contributions to this book show, verbal descriptions of emotional states can provide quite precise information about the specific form of an emotion, such as anger, depression, or happiness, that a person is experiencing.

In addition, there is a range of circumstances under which people talk about emotions that occurred in the past. As Rimé (this volume) shows, people often talk about their past emotional experiences with friends and family. Past experiences are also a major topic of discussion in therapeutic contexts, in self-help groups, and other specialized settings. Furthermore, people talk about others' emotional experiences—people they know, public figures, characters in books and movies, and the like (e.g., Fussell & Moss, 1998). In all these cases, people are communicating about emotions and feelings they are not personally experiencing at the time of the conversation, or at least not experiencing with the same intensity as the original event. Because many nonverbal behaviors are signs rather than intentional signals of emotional state, they have limited value in communicating about emotions one is not experiencing at the time of communication.

Interdisciplinary Approach

Second, the volume takes an explicitly interdisciplinary approach. Valuable insights into the verbal communication of emotion have come from workers in a number of fields, including linguistics, conversational analysis, ethnomethodology, sociolinguistics, anthropological linguistics, communications, and social, cognitive, and clinical psychology (see, e.g., papers in Andersen & Guerrero, 1998; Athanasiadou & Tabakowska, 1998; Niemeier & Dirven, 1997; Russell, 1987). Each of these areas, through its theoretical and empirical approach, offers unique insights into affective communication. The interdisciplinary foundation of the book is evident in several interrelated aspects of the contributions: the level of analysis used to examine verbal phenomena, the authors' empirical approaches, and the context of their investigations.

Multiple Levels of Analysis. The contributors focus on emotional expression at several different levels of analysis. Some focus on specific linguistic devices such as the literal emotional lexicon (e.g., English terms such as *angry, sad, happy* and the like) and/or the use of conventional metaphors, idioms, and other figures of speech (e.g., *hit the roof, down in the dumps, on Cloud 9*). Others examine descriptions of emotions in actual conversations, looking at, among other things, the creation of novel metaphors for emotions. Yet others examine language use at a dialogue level, considering how the emotion is expressed through a series of utterances, looking at the partners' influence, and so forth. Finally, some contributors look at verbal descriptions of emotions over a series of interactions, noting how these descriptions may change with repeated discussion of the emotional incident.

Multiple Empirical Approaches. The contributors also vary in the methodologies they use to approach their subject. The linguistically oriented contributors analyze the meaning and use of conventional expressions for emotions. They consider, for example, how literal and figurative expressions for emotion concepts are expressed in different languages. Other contributors combine quantitative and qualitative analyses of naturally occurring descriptions of emotions, for example, by counting and classifying the number of metaphorical emotion phrases in a dialogue corpus. Contributors with a conversational analytic orientation take a purely qualitative approach, looking closely at how emotions are raised, responded to, and worked through in segments of discourse. Lastly, some contributors take an experimental psychological approach, allowing them to have control over many of the factors hypothesized to influence the production and comprehension of affective language. Each of these approaches has its strengths and weaknesses. By bringing them together in one volume we hope to stimulate greater cross-disciplinary interaction that may lead to converging evidence about the verbal communication of emotions.

Multiple Research Settings. Finally, the contributors focus on how emotions are expressed in a variety of communicative settings. Those taking a linguistic approach consider emotional expressions in the abstract. Others study natural conversations between friends, relatives, and strangers. A number of chapters examine language use in psychotherapeutic contexts, building on previous work by Labov and Fanshel (1977), the contributors in Russell (1987), Siegelman (1990), and others. Finally, some authors pursue their research in the laboratory, where they can carefully control variables such as the number and characteristics of communicators, the topic of conversation, and so forth, to assess the effects of these variables on affective communication.

International Group of Contributors

Finally, the book brings together an international group of contributors. As many of the contributions illustrate, the communication of emotions is shaped by language and culture in a variety of ways. To avoid creating theories that are too heavily rooted in the English language, contributions were solicited from investigators in a number of different countries (Australia, Belgium, Canada, Germany, Hungary, and the United States). Many of these contributors examine affective language in their own and other native languages in addition to English, thereby potentially broadening our scope of understanding.

It should be noted that space limits precluded the inclusion of chapters from every prominent researcher in each of the fields we have mentioned. Each contributor has provided an extensive reference section with pointers to other important research in their respective fields.

OVERVIEW OF CHAPTERS

Chapters are organized into three broad areas: background theory, figurative language use, and social/cultural aspects of emotional communication. Part I, *Theoretical Foundations*, consists of three chapters that look at fundamental issues in the verbal communication of emotion.

Cliff Goddard (chapter 2, *Explicating Emotions Across Languages and Cultures: A Semantic Approach*) discusses a fundamental problem in the study of verbal communication of emotions: semantic differences across languages and cultures. For example, he observes that lexical terms in other languages that are roughly similar to our words *anger* and *depression* can have subtle differences in meaning. As a result, interpreting cross-cultural research on emotional language is problematic. He suggests that instead of glossing over semantic differences between languages, we consider them part of the phenomena to be investigated. Goddard takes an approach known as "natural semantic metalanguage" (NSM), originated by Wierzbicka (1992, 1999). In the NSM approach, word meanings are specified using a small set of universal semantic concepts (e.g., *people*, *good/bad*, *think*, *feel*). Unlike specific emotion terms, he argues, these semantic universals are found in all languages and thus can form metalanguage to describe specific emotion words in specific languages. In applying NSM to emotion terms, feelings associated with a specific emotion (e.g., "sadness") are linked to a typical cognitive scenario (e.g., "something bad has happened") using the semantic metalanguage. Goddard gives a variety of examples of how NSM can be applied to emotion terms in English and a number of other languages including Polish, Malayan, and Japanese. Next, Goddard turns his attention to cultural scripts about expressing emotions. He again applies the NSM strategy of using a small set of universal primitive features to characterize rules for expressing emotions in different cultures. Goddard's chapter is an elegant demonstration of the strengths of the NSM approach.

Sally Planalp and Karen Knie (chapter 3, *Integrating Verbal and Nonverbal Emotion(al) Messages*) focus on how verbal and nonverbal cues to emotion might be theoretically integrated (see also Planalp, 1999). They observe that the complexity of this issue has led to a "divide and conquer" strategy in which investigators tend to focus on individual cues (e.g., facial expressions, intonation, verbal messages) in isolation from the others. Although this strategy has provided insights into emotional communication, it has not increased our understanding of how people integrate nonverbal and verbal cues when expressing and understanding emotions in actual conversations. Planalp and Knie outline O'Keefe's (1988) *Message Design Logics* and explicate the implications of this model for integrating verbal and nonverbal cues to emotion. In *Expressive Logic*, emotions are viewed as entities that build up and escape or leak out of the body in various ways, including nonverbal behaviors, paralinguistic phenomena, and verbal utterances. In *Conventional Logic*, emotional messages are sent, via one or a combination of cues, to a receiver. The focus is on the channels used to send affective messages and the extent to which the recipient understands the message. In *Rhetorical Logic*, emotion and communication are viewed as activities oriented toward the achievement of social goals. Planalp and Knie describe in detail how different conceptualizations of communication affect researchers' choices of topics and paradigms used to investigate emotional communication.

In Chapter 4, *How to do Emotions with Words: Emotionality in Conversations*, Reinhard Fiehler outlines his approach to studying the relationship between emotion and language (Fiehler, 1990), in which emotions are viewed as by nature interactionally-constituted. Fiehler describes three types of emotion rules: *manifestation rules*, which govern the type of emotions and manner in which they are displayed in a particular situation; *correspondence rules*, which specify appropriate responses to others' emotions; and *coding rules,* for identifying instances of emotions in an interaction. He distinguishes among the manifestation of emotions, the interpretation of emotions, and the interactional processing or negotiation of emotions. He also distinguishes between expressions of emotion and their thematization, or explicit verbalization, within the content of the conversation. Fiehler's model is noteworthy in several regards: First, he carefully considers a wide range of ways in which emotions may be communicated, including nonverbal behaviors, paralinguistic phenomena, word selections, conversational dynamics (e.g., interruptions and overlaps), and the like. Second, he describes in detail a number of verbal strategies for thematizing emotions, including verbal labels and descriptions, figurative expressions, and descriptions of circumstances surrounding an emotional event. He discusses how thematizations can focus on different elements in the emotional experience, including the experiencer, type and intensity of emotion, and the dynamics of the experience. Third, Fiehler specifies a six-stage methodology for analyzing the relationship between emotion and communication using his theoretical framework and then analyzes two excerpts from psychotherapy sessions to illustrate the value of his approach. Fiehler's model (explained in detail in his 1990 book) was translated into English specifically for this volume in order to make it more accessible to emotion researchers.

Part II, *Using Figurative Language to Express Emotions*, is comprised of four chapters that look at the role of metaphor and other figures of speech in emotional communication in both everyday language and psychotherapeutic contexts.

In chapter 5 (*Emotion Concepts: Social Constructivism and Cognitive Linguistics*) Zoltán Kövecses sketches out his cognitive linguistic approach to the communication of emotion (e.g., Kövecses, 1996, 2000). He starts by contrasting the cognitive linguistic view of emotion language with that of Harré's "emotionology" (Harré, 1986). He notes that Harré and his fellow social constructivists have tended to focus on literal emotion terms such as *anger, joy,* and *sadness*, and on terms for distinct emotions rather than classes of words for the same emotion (e.g., *angry, irate, irritated*). Kövecses then describes the cognitive linguistic approach, in which particular expressions for emotions are seen as reflecting deeper conceptual structures, which are themselves metaphorical in nature and represent a folk theory of emotion (Lakoff & Johnson, 1980). Kövecses discusses at length, using many examples, how a large number of conventional metaphorical phrases for emotions (e.g., *burst with tears, flipped one's lid*) can be described in terms of an EMOTION AS FORCE conceptual metaphor. He argues that both causes and consequences of emotions are conceptualized metaphorically as forces. Kövecses provides examples showing striking similarities in the conceptual metaphors underlying emotional phrases in different languages and suggests this similarity might stem from similarities in how emotions are experienced in each of these cultures.

Gibbs, Leggitt, and Turner (chapter 6, *What's Special about Figurative Language in Emotional Communication*) also address the role of metaphor and other figures of speech in emotional communication. Gibbs et al. first consider the nature of communicative intentions in metaphor production. They suggest that speakers may use metaphor to convey a variety of subtle meanings, not all of which need have been consciously intended at the time of production. These subtle meanings allow speakers to describe their emotional experiences in more detail than would be possible using terms in the literal emotion lexicon. Like Kövecses, they argue that part of the communicative potential of metaphor stems from the relationship between a particular figurative expression and the deeper conceptual metaphor with which it is associated. Specific emotional metaphors can be seen as reflecting particular phases or aspects of larger conceptual metaphors (e.g., *getting hot under the collar* vs. *hit the roof*), and listeners use this relationship between the two in understanding metaphorical meaning. Gibbs et al. demonstrate in a series of laboratory studies how using different types of figurative language (e.g., irony, metaphor, overstatement, understatement) to express the same class of emotions (e.g., anger) lead to different interpretations about speakers' intentions and emotional states (e.g., how angry the speaker is). Their findings suggest, among other things, that speakers can use figurative language strategically to express subtle nuances of emotional states.

The next two chapters in this part focus specifically on metaphor use in psychotherapeutic contexts. In chapter 7, *Conflict, Coherence, and Change in Brief Psychotherapy: A Metaphor Theme Analysis*, Lynne Angus and Yifaht Korman examine how spoken metaphors and the underlying metaphorical frameworks they represent change over the course of therapy. Angus and Korman suggest that metaphors used in psychotherapy reflect clients' views of themselves, their life circumstances, and the world. Client metaphors may also reflect the nature of the therapeutic relationship. By tracing metaphor use through a series of sessions between a particular client and therapist, they argue, one should be able to identify how clients' views change dynamically over time. To investigate their hypotheses, Angus and Korman analyzed all messages between two clients and their respective therapists over the course of 15 or more sequential sessions. They examined the core metaphorical themes underlying clients' metaphors and other figures of speech as psychotherapeutic progress was made. In particular, they focus on the use of expressions falling under one theme, RELATIONSHIP AS CONFLICT. Angus and Korman found that these metaphors based on this theme could be further classified into three subcategories: FIGHTING AND WINNING, FIGHTING BUT LOSING, and NEGOTIATING. They found that over the course of therapy, the subcategory of metaphorical phrases shifted from FIGHTING BUT LOSING to FIGHTING AND WINNING, although the progression was not linear. The findings highlight the importance of metaphor in therapeutic contexts and suggest that therapists might gain from paying closer attention to clients' metaphors. The findings also provide a nice example of how the language people use to describe specific emotional events can change over time as their understanding of these events evolves through therapy.

In Chapter 8, *Conventional Metaphors for Depression*, Linda McMullen and John Conway focus on metaphors for depression. They first provide a historical perspective on the terms *depression* and *melancholia* and describe the historical bases for several conceptual metaphors for depression. They note that the conceptual metaphors of DEPRESSION IS DARKNESS and DEPRESSION IS WEIGHT date back to ancient Greece. McMullen and Conway draw upon data collected from a study of clients' depression-related metaphors during psychotherapy sessions to show the current pervasiveness of these two metaphors. In addition, they found a third conceptual metaphor, DEPRESSION IS DESCENT, accounted for more than 90% of the depression-related metaphors in their corpus. Clients described themselves as *down, hitting a low, sinking, in the dumps*, and the like. McMullen and Conway go on to show how the DEPRESSION IS DESCENT metaphor fits into a broader up–down spatial framework pervasive in Western cultures, in which "up" is associated with positive properties and "down" is associated with negative properties. Interestingly, they suggest that the very act of speaking about depression via metaphors that fall within the DEPRESSION IS DESCENT conceptual framework, because of its inherent negativity in Western cultures, might serve to worsen client's feelings of worthlessness and despair.

Part III, *Social and Cultural Dimensions*, brings together four chapters that look at ways emotions are embedded in larger sociocultural processes.

Rimé, Corsini, and Herbette (chapter 9, *Emotion, Verbal Expression, and the Social Sharing of Emotion*) look at emotional communication from a somewhat different angle than the previous chapters. Rimé et al. study when and why people choose to share their emotional experiences with others after they have occurred. The stress is on patterns of sharing, rather than on the specific words used to share an experience. Rimé and his collaborators have conducted a large series of studies of the dynamics of social sharing, beginning with studies of people's autobiographical accounts of sharing, and moving to laboratory experiments in which they cleverly manipulate the type of emotional experience a participant undergoes and measure the extent to which this experience is shared with others. A number of notable findings have resulted from this body of work, including the repeated findings that people share the vast majority of their emotional experiences, often very soon after they occur, that they often share these experiences with more than one other person, and that they are especially likely to share more intense emotions. They find that personality characteristics have little effect on the extent of sharing, but that there are gender, age, and culture-related differences in sharing patterns. Also of note was the observed prevalence of secondary sharing, wherein the original recipient of sharing passes the emotion experience on to a third party. Rimé et al. discuss the possible psychological and social functions of sharing emotions, including keeping a community up to date on the experiences of its members, and allowing the sharer to work through his/her experiences and search for meaning for emotional events.

In chapter 10, *The Language of Fear: The Communication of Intergroup Attitudes In Conversations About HIV and AIDS*, Pittam and Gallois examine the language used to express fear, with a focus on language about fears rather than expressions of imminent fear. Their approach (see also Gallois, 1993) is rooted in Tajfel's (1982) Social Identity Theory and other theories of the effects of social category membership on perceptions of oneself and others. Pittam and Gallois observe that the way AIDS has spread among certain social groups (e.g., male homosexuals, intravenous drug users, prostitutes) has made social group memberships especially salient in any discussions about this highly-feared illness. Pittam and Gallois focus on one social category, gender, that they posit is salient in AIDS-related discussions. They present a study of single- and mixed-gender groups discussing the HIV/AIDS situation and safe sex. Content analysis was used to examine all the fear-related words (e.g., *worry, scare, shock, fear*) generated in the discussions. They analyzed the linguistic data in terms of what is feared, who fears, and who is feared, and found that outgroups were often implicitly mentioned in discussions of who was feared. The results illustrate how certain emotional topics of discussion can make participants' social category memberships salient, and how these salient category memberships in turn can affect how people communicate about the topic.

Chapter 11, *Rewards and Risks of Exploring Negative Emotion: An Assimilation Model Account,* by Lara Honos-Webb, Linda Endres, Ayesha Shaikh, Elizabeth Harrick, James Lani, Lynne Knobloch, Michael Surko, and William Stiles, addresses the impact of expressing negative emotions on mental and physical health. They first describe their *Assimilation Model* (Stiles, Elliott, Llewelyn, Firth-Cozens, Margison, Shapiro, & Hardy, 1990), in which clients' representations of negative or problematic experiences are hypothesized to move through a number of stages, from unacknowledged to fully integrated and mastered. Honos-Webb et al. describe two versions of the Assimilation Model, one of which focuses on schema formation and the other of which focuses on how people integrate different "voices." They then present two case studies of assimilation using excerpts from session transcripts to illustrate how clients' affective language shifts as they come to assimilate their experiences. In the next section of the chapter, they report the results of a laboratory study of sharing emotions using a paradigm initiated by Pennebaker (e.g., 1995, 1997). In this study, students wrote for 20 minutes each of four days about their most traumatic emotional experience. In contrast to Pennebaker's findings, the students who had the most assimilation also had the most health care visits in the short but not the long term. Honos-Webb and her colleagues speculate that the four-day writing time might have been long enough to enable participants to uncover previously suppressed or unacknowledged emotional distress, but failed to provide enough time or the proper supportive context for them to work through these issues. Their findings suggest that the expression of traumatic experiences is most beneficial when it occurs in a supportive conversational context wherein people can learn coping strategies to assimilate these experiences.

In Chapter 12, *Blocking Emotions: The Face of Resistance,* Kathleen Ferrara looks at the phenomenon of resistance, or the holding back of emotions, in a psychotherapeutic context. As in her earlier work (Ferrara,

1994), Ferrara uses a discourse-centered approach in which she looks closely at detailed transcripts of actual therapy sessions to determine how both client and therapist jointly construct resistance. She uses Labov and Fanshel's (1977) concepts of *repeated requests*, with which therapists can ask clients to do something, and *put offs*, by which clients can avoid meeting these requests. Ferrara describes discourse rules for making and putting off direct and indirect requests, and shows how components of these requests (e.g., listeners need for the action, need for the request, ability to do the action, willingness to do the action, and so forth) can form the basis of put offs. She provides a number of examples of repeated requests and put offs and then examines the conversational dynamics of a long sequence in which a client repeatedly puts off the therapist's request. The analyses show how the way in which the therapist phrases the request (indirectly) enables the client to continue his series of put offs. Ferrara suggests that direct requests might lead to less resistance on the part of clients. Her work demonstrates how one's communicative partner can influence the ways in which emotions are expressed or (or fail to be expressed). She argues that by better understanding how clients and therapists co-construct conversation, therapists should be able to provide more effective therapy.

THEMES AND ISSUES FOR FUTURE RESEARCH

As can be seen from these brief synopses, contributors to this volume address a variety of issues in the verbal communication of emotion. Taken together, the chapters provide substantial insights into how people talk about affective experiences. In this section I outline some broad questions that arise from this body of work and some areas for future research.

What Is the Nature of the Relationship Between Verbal and Nonverbal Communication of Emotion?

Although contributions to this volume were limited to those focusing predominantly on the *verbal* communication of emotion, many chapters touch on the impact of nonverbal and paralinguistic cues as well. This can be seen, for example, in the therapy session excerpts provided by Fiehler, Honos-Webb and colleagues, and Ferrara. In each case, nonverbal signs such as sighs, laughs, and the like, are incorporated into the written transcript. It seems clear that we, as readers of these transcripts, incorporate the transcribed nonverbal signs into our understanding of the emotional dynamics of the sessions. As Goddard and Fiehler argue, our interpretations of nonverbal signs may be shaped by cultural scripts specifying what behaviors are expected in particular settings. Quite a bit is known about the role of different nonverbal and paralinguistic cues to emotion (cf. Ekman & Davidson, 1994; Feldman & Rimé, 1991; Philippot, Feldman, & Coats, 1999; Russell & Fernandez-Dols,

1997; Scherer & Ekman, 1984). Yet, as Planalp and Knie point out, the ways in which we integrate nonverbal cues with verbal utterances is still poorly understood. Their chapter provides a number of suggestions for future research in this area.

An understanding of the ways verbal and nonverbal cues are integrated has become especially relevant today, now that new technologies allow for communication via text-based chat, e-mail, and other media in which verbal communication is the primary channel of communication (e.g., Kiesler, Siegel, McGuire, 1984; Rice & Love, 1987; Walther & Burgoon, 1992). The use of emoticons such as smiley faces to indicate nonverbal cues to emotion in text-based media suggests that people believe that nonverbal cues are an important addition to their words when expressing emotion. A better understanding of how people integrate verbal and nonverbal cues in face-to-face settings would enable system designers to develop technologies to support emotional communication among remotely distributed parties.

Two of the chapters suggest that the relationship between verbal and nonverbal cues to affect may run deeper than their simultaneous use to express an ongoing emotional state. Both Kövecses and Gibbs et al. provide examples showing that many metaphors for emotional states were originally rooted in bodily experiences (e.g., *hot under the collar*). Kövecses suggests that the similarity of conceptual metaphors across cultures may stem from the universality of the bodily components of emotional responses (e.g, getting flushed). The question thus arises as to whether the actual experiencing of nonverbal signs of emotional states influences the figurative expressions speakers choose to communicate those states.

What Are the Relationships Among Felt Emotions, Cognitions, and Affective Language?

In part, an answer to the question of how verbal and nonverbal cues to emotion are integrated may depend on the nature of the relationship between felt emotions and their verbal communication. In some cases people talk about emotional experiences as they are happening, whereas at other times they relate their own and others' past emotional experiences (Rimé et al.). The degree to which the original emotion is felt at the time of sharing may thus vary substantially. Because nonverbal cues to emotion are often a by-product of the actual experience of an emotion, the ways these cues are integrated with verbal messages may differ depending on the extent to which the emotion is felt at the time of the utterance. Another, almost Whorfian (Whorf, 1956) angle on the emotion/language relationship is suggested at least indirectly by some of the chapters. For example, McMullen and Conway suggest that by speaking about their emotions using the DEPRESSION IS DESCENT conceptual metaphor, people's feelings of sadness might worsen. Thus, there may be a bidirectional relationship between felt emotions and the verbal expression of these emotions that requires further investigation.

Likewise, a bidirectional relationship between cognition and emotional language is suggested by the chapters on social sharing of emotions (Rimé et al.), resistance (Ferrara), and assimilation (Honos-Web et al.). In each of these cases, it appears that expressing emotions to others has, in at least some cases, positive benefits for the speaker's cognitive views of the emotional circumstance. Honos-Webb and colleagues show, for example, how in a therapeutic context descriptions of an emotional experience come to reflect clients' greater assimilation of that experience. Angus and Korman show how changes over time in clients' selections of metaphors from the RELATIONSHIPS ARE CONFLICT domain reflect changes in their conceptualizations of their relationships. Furthermore, Ferrara and Rimé et al. suggest that failures to share emotions with others can, at least in some cases, lead to failure to resolve negative experiences. Thus, another direction for future research would be to expand studies of the relationships between affect and cognition (see, e.g., papers in Clark & Fiske, 1982; Eich, Kihlstrom, Bower, & Forgas, 2000; Fiedler & Forgas, 1988; Izard, Kagan, & Zajonc, 1984) to the examination of relationships among emotions, cognitions, and verbal affective communication.

To What Extent Are Emotions and Affective Communication Co-constructed by Speaking Partners?

In the Rhetorical model described by Planalp and Knie, affective communication is embedded in a social context that can shape its form, and this theme is addressed in a number of the contributions to this volume. Honos-Webb and colleagues, for example, suggest that the benefits of talking about emotions in their case studies stem in part from clients' ability to work the issues through with their therapist. Rimé et al. also suggest that the sharing of emotions has benefits by enabling people to work through their experiences. Conversational grounding (e.g., Clark, 1996) and perspective-taking (cf. Krauss & Fussell, 1996) models of communication likewise suggest that how people formulate messages about emotional states will be influenced by their partners' characteristics and responses.

One's communicative partner can influence affective communication in a number of ways. First, Rimé points out that others' characteristics (particularly their relationship with the speaker) is a determinant of with whom emotions will be shared. In addition, as Pittam and Gallois' contribution highlights, a partner's social category memberships can make the speaker more aware of his or her own category memberships and thereby influence the way that an emotion is discussed. Finally, Ferrara shows here and in her earlier book (Ferrara, 1994) that conversational partners' responses can impact how emotional messages are formulated.

A better understanding of the effects of conversational partners on affective communication is thus another area for future research, especially in the light of the previous section suggesting that there may be bidirectional relationships between how emotions are expressed and people's experiences of emotion and cognitions about emotional events.

How Does Culture Influence Emotional Experience and Affective Communication?

Most of the chapters touch at least indirectly on the role of sociocultural processes in emotional communication. However, the impact of culture on emotional language remains somewhat unclear. On the one hand, several chapters describe cultural differences. For example, Goddard discusses the cultural embeddedness of the emotion lexicon as well as the role of culture in displaying emotions; Fiehler notes that there are cultural differences in rules for displaying emotions, responding to others' emotional displays, and interpreting emotions; Rimé and colleagues found differences in patterns of sharing emotions between Western and Eastern cultures; and McMullen and Conway suggest that metaphors for depression are based in Western conceptualizations of the self. On the other hand, Kövecses argues that many conceptual metaphors for emotion are universal, perhaps because they are rooted in bodily manifestations of emotional experiences that are likely to be fairly constant across cultures. The ways in which culture shapes emotional expression are likely to be complex (cf. Kitayama & Markus, 1994) and require further study.

CONCLUSION

This volume brings together an international, interdisciplinary group of researchers focusing on the verbal communication of emotion. The contributions illustrate the value of the authors' respective theoretical and empirical approaches for an understanding of affective language in both everyday and clinical settings. Taken as a whole, the chapters provide a comprehensive look at the current state of research on the use of language in affective communication and suggest a number of interesting directions for future research.

REFERENCES

Andersen, P. A., & Guerrero, L. K. (Eds.). (1998). *Handbook of communication and emotion: Research, theory, applications, and contexts*. San Diego, CA, Academic Press.

Angus, L. E. (1996). An intensive analysis of metaphor themes in psychotherapy. In J. S. Mio & A. N. Katz (Eds.), *Metaphor: Implications and applications* (pp. 73–84). Mahwah, NJ: Lawrence Erlbaum Associates.

Athanasiadou, A., & Tabakowska, E. (Eds.). (1998). *Speaking of emotions: Conceptualization and expression*. The Hague: Mouton de Gruyter.

Barrett, K. (Ed.). (1998). *The communication of emotion: Current research from diverse perspectives*. San Francisco: Jossey-Bass.

Bush, L. E. (1973). Individual differences multidimensional scaling of adjectives denoting feelings. *Journal of Personality and Social Psychology, 25*, 50–57.

Clark, H. H. (1996). *Using language*. Cambridge: Cambridge University Press.

Clark, M. S., & Fiske, S. T. (Eds.). (1982). *Affect and cognition*. Hillsdale, NJ: Lawrence Erlbaum Associates.

Clore, G. L., Ortony, A., & Foss, M. A. (1987). The psychological foundations of the affective lexicon. *Journal of Personality and Social Psychology, 53*, 751–766.

Davitz, J. R. (1969). *The language of emotion*. New York: Academic Press.

Davitz, J. R., & Mattis, S. (1964). The communication of emotional meaning by metaphor. In J. R. Davitz (Ed.), *The communication of emotional meaning* (pp. 157–176). Westport, CT: Greenwood Press.

Eich, E., Kihlstrom, J. F., Bower, G. H., & Forgas, J. (Eds.). (2000). *Cognition and emotion*. Oxford: Oxford University Press.

Ekman, P., & Davidson, R. J. (Eds.). (1994). *The nature of emotion: Fundamental questions*. New York: Oxford University Press.

Feldman, R. S., & Rimé, B. (Eds.). (1991). *Fundamentals of nonverbal behavior*. Cambridge: Cambridge University Press.

Ferrara, K. (1994). *Therapeutic ways with words*. New York: Oxford University Press.

Fiedler, K., & Forgas, J. (Eds.). (1988). *Affect, cognition and social behavior*. Toronto: C. J. Hogrefe.

Fiehler, R. (1990). *Kommunikation und emotion: Theoretische und empirsche untersuchungen zur Rolle von Emotionen in der verbalen Interaktion*. [Communication and emotion: Theoretical and empirical investigations of the role of emotions in verbal interaction.] Berlin: Walter de Gruyter.

Fussell, S. R., & Moss, M. M. (1998). Figurative language in emotional communication. In S. R. Fussell & R. J. Kreuz (Eds.) *Social and cognitive approaches to interpersonal communication* (pp. 113-141). Mahwah, NJ: Lawrence Erlbaum Associations.

Gallois, C. (1993). The language and communication of emotion: Universal, interpersonal, or intergroup? *American Behavioral Scientist, 36*, 262-270.

Gibbs, R. W. Jr. (1994). *The poetics of mind: Figurative thought, language, and understanding*. New York: Cambridge University Press.

Harré, R. (1986). An outline of the social constructionist viewpoint. In R. Harré (Ed.), *The social construction of emotion*. Oxford: Basil Blackwell.

Izard, C. E., Kagan, J., & Zajonc, R. B. (Eds.) (1984). *Emotions, cognition and behavior*. Cambridge: Cambridge University Press.

Johnson-Laird, P. N., & Oatley, K. (1989). The language of emotions: An analysis of a semantic field. *Cognition and Emotion, 3*, 81–123.

Karp, D. A. (1996). *Speaking of sadness: Depression, disconnection, and the meanings of illness*. New York: Oxford University Press.

Kiesler, S., Siegel, J., & McGuire, T. W. (1984). Social psychological aspects of computer-mediated communication. *American Psychologist, 39*, 1123-1134.

Kitayama, S., & Markus, H. (1994). *Emotion and culture: Empirical studies of mutual influence*. Washington, DC: American Psychological Association.

Kövecses, Z. (1996). *Metaphors of anger, pride and love: A lexical approach to the structure of concepts*. Philadelphia: John Benjamins.

Kövecses, Z. (2000). *Metaphor and emotion: Language, culture, and body in human feeling*. Cambridge: Cambridge University Press.

Krauss, R. M., & Fussell, S. R. (1996). Social psychological models of interpersonal communication. In E. T. Higgins & A. Kruglanski (Eds.), *Social psychology: Handbook of basic principles* (pp. 655-701). NY: Guilford Press.

Labov, W., & Fanshel, D. (1977). *Therapeutic discourse: Psychotherapy as conversation*. New York: Academic Press.

Lakoff, G., & Johnson, M. (1980). *Metaphors we live by*. Chicago: University of Chicago Press.

McMullen, L., & Conway, J. (1996). Conceptualizing the figurative expressions of psychotherapy clients. In J. S. Mio & A. N. Katz (Eds.), *Metaphor: Implications and applications* (pp. 59–71). Mahwah, NJ: Lawrence Erlbaum Associates.

Niemeier, S., & Dirven, R. (Eds.) (1997). *The language of emotions. Conceptualization, expression, and theoretical foundation*. Amsterdam: John Benjamins.

O'Keefe, B. J. (1988). The logic of message design: Individual differences in reasoning about communication. *Communication Monographs, 55*, 80-103.

Ortony, A., Clore, G. L., & Collins, A. (1988). *The cognitive structure of emotions*. Cambridge: Cambridge University Press.

Ortony, A., Clore, G. L., & Foss, M. A. (1987). The referential structure of the affective lexicon. *Cognitive Science, 11*, 341–364.

Pennebaker, J. W. (Ed.). (1995). *Emotion, disclosure, and health*. Washington, DC: American Psychological Association.

Pennebaker, J. W. (1997). Writing about emotional experiences as a therapeutic process. *Psychological Science, 8*, 162-166.

Philippot, P., Feldman, R. S., Coats, E. J. (Eds.). (1999). *The social context of nonverbal behavior*. Cambridge: Cambridge University Press.

Planalp, S. (1999). *Communicating emotion: Social, moral, and cultural processes*. Cambridge: Cambridge University Press.

Pollio, H., & Barlow, J. (1975). A behavioral analysis of figurative language in psychotherapy: One session in a single case study. *Language and Speech, 18*, 236–254.

Rice, R. E., & Love, G. (1987). Electronic emotion: socio-emotional content in a computer-mediated communication network. *Communication Research, 14*, 85-105.

Russell, J. A., & Fernandez-Dols, J-M. (Eds.). (1997). *The psychology of facial expression*. Cambridge: Cambridge University Press.

Russell, R. (Ed.). (1987). *Language in psychotherapy: Strategies of discovery*. New York: Plenum.

Scherer, K., & Ekman, P. (1984). *Approaches to emotion*. Hillsdale, NJ: Lawrence Erlbaum Associates.

Siegelman, E. (1990). *Metaphor and meaning in psychotherapy*. New York: Guilford.

Stiles, W. B., Elliott, R., Llewelyn, S. P., Firth-Cozens, J. A., Margison, F. R., Shapiro, D. A., & Hardy, G. (1990). Assimilation of problematic experiences by clients in psychotherapy. *Psychotherapy, 27*, 411-420.

Tajfel, H. (1982). *Social identity and intergroup relations*. Cambridge: Cambridge University Press.

Walther, J. B., & Burgoon, J. K. (1992). Relational communication in computer-mediated interaction. *Human communication research, 19*, 50-88.

Whorf, B. L. (1956). (J. B. Carroll, Ed.) *Language, thought and reality: Selected writings of Benjamin Lee Whorf*. Cambridge, MA: MIT Press.

Wierzbicka, A. (1992). *Semantics, culture, and cognition*. Oxford: Oxford University Press.

Wierzbicka, A. (1999). *Emotions across languages and cultures: Diversity and universals*. Cambridge: Cambridge University Press.

PART I

Theoretical Foundations

– 2 –

Explicating Emotions Across Languages and Cultures: A Semantic Approach

Cliff Goddard
University of New England, Australia

EMOTIONS: THE PERSPECTIVE FROM CROSS-CULTURAL SEMANTICS

Cross-cultural research of any kind cannot afford to ignore the problems posed by semantic differences between languages. These problems are particularly pertinent for psychology, given that information about other people's mental states is inevitably mediated by language. Unfortunately, however, social scientists often regard the problem of translation as a mere methodological nuisance—as something to be "gotten around" so that they can move on to implementing familiar research techniques, rather than as a profound epistemological and conceptual issue deserving of sustained and focused attention. At the same time they underestimate both the scope of semantic variation between ethnopsychological lexicons, and the hazards of uncritically using English as the metalanguage of cross-cultural description.

This tendency is evident even among anthropologists, who are more sensitive than most to the possibility of deep-rooted conceptual differences between languages and cultures. As Lutz (1988) commented, it is a shibboleth of anthropological method that "all ethnography is comparative, involving the implicit or explicit comparison of the culture of the observer with that of the observed" but in practice this "rarely leads to the simultaneous examination of both meaning systems" (p. 44). Instead, indigenous emotion concepts are merely "tagged" with English glosses. This practice not only brings with it an obvious danger of ethnocentric distortion, it also excuses the analyst from engaging in deep conceptual analysis of English folk categories, which continue to be mistaken for objective categories of psychological reality.

Discussions of this issue in anthropology, psychology and sociology often get bogged down in fruitless exchanges between entrenched relativist and universalist positions. My contention is that recent developments in linguistic semantics have something new and useful to contribute, which can help overcome some of the conceptual confusions and methodological difficulties impeding the development of a soundly based cross-cultural psychology. I refer to the "natural semantic metalanguage" (NSM) approach to semantic analysis of Anna Wierzbicka and colleagues (cf. Goddard, 1998; Goddard & Wierzbicka, 1994, in press; Wierzbicka, 1992a, 1996a).[1] In this chapter I explain and demonstrate the value of this approach, but before that some further elaboration is in order on the hazards of not taking semantic differences between languages seriously enough.

Researchers in cultural psychology and ethnopsychology (e.g. Harré, 1986; Lutz, 1988; Shweder, 1991, 1993; White & Kirkpatrick, 1985) are usually highly sensitive to the difficulties of translation, but unfortunately the same can hardly be said of psychology at large. For example, in an edited volume titled *The Nature of Emotion*, with the subtitle *Fundamental Questions* (Ekman & Davidson, 1994), the terms "translation" and "language" do not even appear in the index; and of over 20 contributors, only one identifies language issues as posing any great conceptual or methodological dilemma. The exception, not surprisingly, is Richard Shweder, who, referring to "the all too often glossed over problem of translation" said that: "the range of implications, suggestions and connotations of psychological state terms do not easily map, at least not lexically, from one culture to another" (1994, p. 33). When language and translation issues are mentioned (in this book, and elsewhere) it is more usual to find statements along these lines: "[e]quivalents for most of the emotions terms commonly considered part of the 'basic' category seem to exist in almost all major languages of the world" (Scherer, 1994, p. 25).

Leaving aside the question of what counts as a "major language" for the purposes of psychology, it is instructive to have a closer look at how translation is handled in large intercultural studies of emotion, such as those conducted by Klaus Scherer and his collaborators. Scherer, Wallbott, and Summerfield (1986) administered a questionnaire in eight European countries with the aim of assessing the frequency and quality of emotional experience. Information was sought on four supposedly universal categories, each of which was characterized by two words: *joy/happiness, sadness/grief, fear/fright, anger/rage*. Aebischer and Wallbott (1986) explain that, "[T]hese labels were first decided on in the English language, then translated into the respective other languages, and finally back-translated to guarantee equivalence across languages. Thus we tried to be sufficiently precise about the kind of emotion to be evoked" (p. 32). But can the

---

[1]The NSM bibliography is extensive and cannot be reviewed her for reasons of space. Aside from works cited elsewhere in the chapter, representative linguistic works include: Ameka (1990), Chappell (1986), Peeters (1993), and Wilkins (1986).

procedure of back-translation really guarantee equivalence of meaning? Hardly—it merely assures us that the terms being used are the closest single-word equivalents (or near-equivalents) available in the various languages.

As a matter of fact, even among the languages of Europe there are significant differences in emotion semantics. For example, in the German version of the questionnaire, *Angst* was one of the words used for the item *fear/fright*, but it is well known that the meaning of *Angst* is quite different from that of either English *fear* or *fright*. More subtle, but just as real, are differences between French *tristesse* and English *sadness*, between German *Glück* and English *happiness*, between Italian *rabbia* and English *anger* (or *rage*), and so on (cf. Wierzbicka, 1992b, 1993). In reality, the survey instrument was not comparing like with like across the range of languages concerned, which inevitably compromised the validity of the results (to an unknown extent).

Translation problems are compounded when respondents have to work in the "forced choice" strategy used in many experiments on the so-called "recognition" of facial expressions. As argued by Russell (1991), the most such experiments can show is that people in different cultures give similar (but not necessarily identical) interpretations to facial expressions. To see why, said Russell, imagine that you are a subject asked to select one of the six terms to describe a face wearing a bright smile:

> Most likely, you'd select *happy*. But now suppose that *happy* had been replaced on the list with *elated*. Given the alternatives, you'd have no choice but to select *elated*. If *happy* were successively replaced with *serene, satisfied, excited, grateful*, and *triumphant*, you'd again select any of these words in turn...Indeed, substitute for *happy* any clearly positive word...and the conclusion remains the same. (p. 435)

Despite this obvious weakness, such experiments are often referred to in the secondary literature as proof that all languages have words for the basic emotions.

Of course, aside from the "recognition" experiments, many other techniques (similarity judgment tasks, sorting tasks, listing tasks, rating tasks, etc.) have been used by anthropologists and cross-cultural psychologists. It would be impossible to comment on all of them here, but no matter how useful these various techniques may be, the point remains that none of them escapes the problem of translation as they all depend on elicitation and/or interpretation in natural language.

Often psychologists dismiss semantic differences between languages as matters of connotation, cultural emphasis, and so on. What does it matter if there is no exact equivalent to English *anger* or to English *fear* in some exotic language? The true issue, they say, is not a question of word meanings but of real psychological processes. In my opinion, responses like these underestimate the force of the charge that test instruments are flawed by inaccurate translation.

They also bring to light conceptual issues concerning the mission and underlying assumptions of psychology.

One reason why the meanings of emotion words is of fundamental importance to psychology is that, as Fehr and Russell (1984) put it: "Part of the psychologist's job in such cases is to understand emotion concepts as people use them in everyday life" (p. 483). In this light, semantic differences between emotion lexicons should not be viewed as a mere nuisance but as part of the phenomena to be investigated. It is surely incontestable that the words and phrases that people use on a daily basis to describe and to negotiate their emotional experiences are an invaluable key to everyday conceptualization (folk psychology). Equally, as scholars have often observed, language plays a key role in an individual's socialization and thus has a formative influence on the constitution of the individual psyche. As Jerome Bruner (1990) has remarked: "We learn our folk psychology early, learn it as we learn to use the very language we acquire and to conduct the interpersonal transactions required in communal life" (p. 35).

A second reason why the meanings of emotion words should be of fundamental concern is suggested by Edward Sapir's warning that "the philosopher [and one might add, the psychologist] needs to understand language if only to protect himself against his own language habits" (1949, p. 165). Sapir is referring to the tendency for people to assume that the folk taxonomy embedded in their mother tongue represents a reliable guide to an independently existing, objective reality. As readers of this book will be aware, it can be argued that some of the foundational postulates of Western psychological theory have been derived, unwittingly, from language-specific peculiarities of European languages. For example, it has been argued that traditional categories such as "emotion," "sensation," and "cognition," and dichotomies such as "mind" vs. "body," are but cultural artifacts of a certain Western intellectual tradition; and that they need to be deconstructed—and reconstructed—if psychology is to have any chance of transcending shallow ethnocentrism (cf., among others, Harré, 1986; Lutz, 1988, Wierzbicka, 1993). Similarly, it can be argued that using culture-bound terms like *sadness*, *anger*, and *fear* as labels for putatively objective psychological universals is profoundly confused and ethnocentric.

In an effort to come to grips with such problems, some scholars have attempted to recast their formulations into terms that they hope are more language-neutral or more universal. For example, Ortony, Clore, and Collins (1988) endeavored to characterize the "cognitive eliciting conditions" of a set of putative universal emotion-types using terms that were "as language-neutral... as possible"; the emotion type "anger," for instance, was characterized by the conditions "disapproving of someone's blameworthy action" and "being displeased about an undesirable event". But words like *disapprove*, *displeased*, *blameworthy*, and so on, are just as language-specific as *anger* itself, and encapsulate complex language-specific concepts. On another tack, scholars such as Lazarus (1991, 1995) have attempted to link a set of putative universal emotional reactions with a set of universal "core relational themes"; for example, *sadness* is linked with an experience of "irrevocable loss" and *anger* with a "demeaning offense." Unfortunately, however, concepts like *loss* and *offense* are no more universal than are *sadness* and *anger*.

At this point, many scholars begin to experience frustration and impatience. They don't want to be dragged into interminable debates about the meanings of words, and they don't want to succumb to extreme relativism either. A common response is to declare that one is not interested in minute or precise details: that terms like *sadness* or *anger*, when used in a technical sense, are not intended to stand for the same meanings that they have in ordinary English but instead denote "families" of similar emotions. For example, Plutchik (1994) used the term *joy* to subsume also *love, pleasure, elation, happiness,* and *satisfaction,* all of which he regarded (incredibly!) as "near equivalents." Ekman (1992, 1994) has recently enunciated a concept of "emotion families": each emotion (as Ekman now uses the term) "is not a single affective state, but a family of related states... [which] can be considered to constitute a theme and variations" (1994, p. 19). However, this does not really get around the basic problem. One must still find a way to specify the common themes (or components) of the emotion families in a clear and language-neutral fashion.

In short, psychology has no real option but to engage with the problem of cross-linguistic semantic variation in a more sustained and focused way. Like Russell (1991), I believe that this not only calls for more evidence, but also for "new methods." In this respect, linguistic semantics has much to offer.

Among the various schools of linguistic semantics, the one which has devoted the most attention to cross-cultural semantics is the "natural semantic metalanguage" (NSM) framework (see references cited earlier). In this framework, the meaning of a word is stated in the form of an explanatory paraphrase composed in a small, standardized and translatable metalanguage based on natural language. The lexicon of this semantic metalanguage consists of so-called "semantic primes," which research indicates have exact exponents in all or most of the world's languages. Examples include: *someone/person, people, good, bad, think, know, want, feel, do, happen.*

The status of words like these is quite different to that of words like *emotion, sadness,* and *disgust.* We know that many languages (even European languages like German and Russian) lack exact semantic equivalents to the English term *emotion*; that many languages (e.g., Tahitian) lack exact equivalents to *sadness*; that many languages (e.g., Polish) lack equivalents to *disgust* (cf. Levy, 1973; Wierzbicka, 1992a, 1993, 1994b). But no language is yet known that lacks a discrete expression (be it a separate word, an affix, or a fixed phrase) for meanings like *think, know, want,* or *feel*; or for meanings like *people, good, bad*, *do*, and *happen*. Thus, if we can frame our analyses of emotion meanings in terms of semantic primes, we can avoid imposing a language-specific interpretation in the very terms of description.

To forestall possible misunderstandings, it should be pointed out that claiming that all languages have a word for *good* is quite different to claiming that people in every culture have the same ideas about what is *good* (which obviously they do not). Likewise, claiming that all languages have a word for *people* does not mean that everyone has the same ideas about human nature. It should also be added that identifying exponents of semantic primes across languages is complicated by various factors, especially by differences in secondary (i.e., polysemic) meanings. For example, English *want* not only has its primitive sense (which, for typographical clarity, may be written as WANT), but also a secondary sense approximating "need" (as in *This floor wants cleaning*). The Spanish exponent of WANT *querer* has a different secondary sense, which approximates English "love, like" (as in *El me quiere* "he likes me"). Cases like these can be confusing in the absence of careful linguistic analysis.

At present, the lexicon of the NSM metalanguage numbers about 60 to 65 elements. Along with a universal set of simple lexicalized meanings, the NSM theory also posits a universal set of patterns according to which the primitive meanings may be combined, so that, effectively, there are certain phrases and sentences that can be transposed from one language to another with no change of meaning. Readers who wish to pursue these issues are referred to Goddard and Wierzbicka (1994), Wierzbicka (1996a), and Goddard and Wierzbicka (in press).

The NSM method has been applied extensively to emotion semantics. Wierzbicka's work includes examples from Polish, Russian, Hawaiian, Tahitian, and Ifaluk (Wierzbicka 1992a, 1992b, 1992c, 1994a, 1994d, 1995, 1999), as well as English. Others have done work on Yankunytjatjara and Malay (Goddard, 1991, 1995, 1996, 1997a), Aboriginal English and Maori (Harkins, 1990, 1996), and Japanese (Hasada, 1994, 1998; Travis, 1998), among other languages (see especially Harkins & Wierzbicka, in press).

THE SEMANTICS OF EMOTION TERMS: EXAMPLES FROM ENGLISH

The main idea of the NSM approach to emotion terms is that the meaning of an emotion term involves reference to a feeling that is linked with a characteristic (prototypical) cognitive scenario. Wierzbicka has employed this concept from her earliest work on emotions, taking her cue from Tolstoy's practice of suggesting subtle emotions by means of ingenious hypothetical scenarios. Consider these two examples from *Anna Karenina*.

From after dinner till early evening, Kitty felt as a young man does before a battle.

He [Karenin] felt now rather as a man might do returning home and finding his own house locked up.

References to imaginary situations of this kind, although highly evocative, are essentially individual and don't have the force of generalizations. Wierzbicka's insight was that the emotion words of ordinary language work in a similar fashion, except that instead of linking feelings with illustrative situations they link them with cognitive scenarios involving thoughts and wants. For example, and very roughly, *sadness* is a bad feeling linked with the thought "something bad happened," *remorse* is a bad feeling linked with the thought "I did something bad," *joy* is a good feeling linked with the thought "something very good is happening now." Of course, it is quite possible to feel some emotions (for instance, *sadness* or *happiness*) without being aware of the cause, but this fact is fully compatible with an analysis based on a prototypical scenario. An analysis of *joy*, for instance, which links it with the thought "something very good is happening now," does not insist that every time one feels *joy*, one necessarily thinks this particular thought. Rather, it says that to feel *joy* is to feel like someone would who is thinking that thought. The scenario serves as a kind of "reference situation" by which the nature of the associated feeling can be identified.

Explications of this kind are able to capture very subtle differences between the meanings of related emotion terms in a single language, and of similar but non-identical emotion terms across languages. Just as importantly, because such explications are readily intelligible and translatable, they can be tested both directly, against native speakers' intuitions, and indirectly, by checking that they account for the attested range of use of the words in question.

The following analyses are drawn from Wierzbicka (1999).

"Sad" versus "Unhappy"

When the concept of "sadness" has been discussed in the literature, scholars generally link it with "loss." For example, Harris (1989, p. 103) linked "sadness" with the situation "when desirable goals are lost"; Lazarus (1991, p. 122) assigned it the "core relational theme" of "having experienced an irrevocable loss." Harris' analysis allows him to capture some relationships between *sadness* and certain other emotion concepts, notably *anger* ("desirable goals blocked") and *joy* ("desirable goals achieved"), but it does not capture the similarities and differences between *sadness* and, for example, *unhappiness*, *distress*, or *disappointment*. It is also inconsistent with linguistic evidence, because the term *goal* implies that one is doing something because one wants something to happen whereas the word *sad* can be applied to situations where no goals are involved at all. For example, I may feel sad when I hear that my friend's dog died, but this has nothing to do with any goals that I may have had.

Lazarus' suggestion that *sadness* is linked with an "irrevocable loss" is not sustainable either. If anything, "irrevocable loss" is linked with *grief* rather than with *sadness*, which doesn't have to be linked with personal losses at all. In support of this point, Wierzbicka (1999) adduced the following statement by a woman visiting a colleague in hospital who is dying of cancer: *I miss you a lot at work (...) I feel so sad about what's happening to you.* The cause of the visitor's sadness is not the fact that she is losing a colleague, but rather the "bad thing" that has happened (the colleague's illness) and her awareness that she can't do anything about it.

The prototypical cognitive scenario associated with the concept *sad* involves an awareness that "something bad has happened" (not necessarily to me) and an acceptance of the fact that one can't do anything about it. The full explication for the English word *sad* can be represented more precisely as follows:

1. *Sad* (X was sad) =
   - (a) X felt something
   - (b)   sometimes a person thinks:
   - (c)   "I know: something bad happened
   - (d)   I don't want things like this to happen
   - (e)   I can't think: I will do something because of it now
   - (f)   I know I can't do anything"
   - (g)   when this person thinks this, this person feels something bad
   - (h) X feels something like this

As has often been pointed out, a person who feels *sad* may not be conscious of the reason for the sadness, and one can say *I feel sad today, I don't know why.* But this is perfectly consistent with explication (1). The phrasing of the explication allows that a person who is said to feel *sad* doesn't have to think about anything in particular. Nonetheless, the feeling of sadness can only be characterized by reference to a prototypical scenario that does involve a particular set of thoughts.

Unhappiness differs from *sadness* in a number of ways. To begin with, it does require the experiencer to have some certain real thoughts. For although one can say *I feel sad, I don't know why*, it would be a little odd to say *I feel unhappy, I don't know why.* Second, *unhappy* implies a more intense feeling and a stronger negative evaluation, as implied by the fact that it is less readily combinable with minimizing qualifiers like *a little* or *slightly.* Compare (a question-mark in front of an expression indicates that there is something odd or anomalous about it):

> She felt a little (slightly) sad.
> ?She felt a little (slightly) unhappy.

Third, *unhappy* has a more personal character than *sad*: One can be saddened by bad things that have happened to other people, but if one is unhappy, it is because of bad things that have happened to one personally. Fourth, *unhappy*—in contrast to *sad*—does not suggest a resigned state of mind. If in the case of *sadness* the experiencer focuses on the thought "I can't do anything about it," in the case of *unhappiness* he or she focuses on some thwarted desires ("I wanted things like this not to happen to me"). The attitude is not exactly active because one doesn't necessarily want anything to happen, but it is not passive either. Finally, *unhappy* seems to suggest, prototypically, a state extended in time rather than a momentary occurrence (cf. *a moment of sadness* vs. ?*a moment of unhappiness*).

2. *Unhappy* (X was unhappy) =
   (a) X felt something because X thought something
   (b)   sometimes a person thinks for some time:
   (c)   "some very bad things happened to me
   (d)   I wanted things like this not to happen to me
   (e)   I can't not think about it"
   (f)   when this person thinks this, this person feels something bad for some time
   (g) X felt something like this
   (h) because X thought something like this

"Outraged" versus "Appalled"

As indicated by the following quotation, being *outraged* and *appalled* have much in common. Basically, in both cases the (prototypical) experiencer thinks that "something very bad happened" and that one wouldn't have thought something like this could happen.

> When allegations of physical and sexual violence emanate from a classroom, parents are *outraged*, the community *appalled*.

Nonetheless, there are differences as well as similarities between the two concepts. The challenge, for linguistic semantics, is to articulate both the similarities and differences in clear and testable terms.

In some ways, being *outraged* is similar to being *indignant*. As well as thinking that someone did something very bad and unexpected, the experiencer takes an attitude that is at once unaccepting and protesting (along the lines of "I don't want things like this to happen") and at the same time, active and goal oriented ("I want to do something because of this"). In addition, *outrage* implies a thought not only that "someone did something very bad," but also that as a result "something very bad happened." This last factor may account, to some extent, for the greater moral weight of *outrage* (not shared, e.g., by *indignant*, which lacks this last component).

Another factor may be the "social character" of *outrage*: the perpetrators must be, or represent, a group, and so must the victims, or those who identify with the victims. For example, if someone mistreats my child I would be *angry*, even *furious*, rather than *outraged*. On the other hand, if I discover that some teachers mistreat children in the school my child attends then I could indeed be *outraged*. At the same time, I cannot be *outraged* over the treatment of children in Uganda or Sudan (although I can be *shocked* and *appalled* by it). To be *outraged*, it seems, I have to have a role in the situation, a role that makes it imperative for me to take an interest in the matter and to act on behalf of other people.

This leads to the following explication:

3. *Outrage* (X was outraged) =
   (a)   X felt something because X thought something
   (b)   sometimes a person thinks:
   (c)   "I know now: something very bad happened to some people
   (d)   because some other people did something very bad
   (e)   I didn't think that these people could do something like this
   (f)   I don't want things like this to happen
   (g)   I want to do something because of this"
   (h)   when this person thinks this, this person feels something bad
   (i)   X felt something like this
   (j)   because X thought something like this

Now to *appalled*. The example sentence quoted above is suggestive of some differences in attitude between *outraged* and *appalled*. The parents feel responsible for their children; they want to do something about the situation and they think they have to do so. The community, on the other hand, reacts more as onlookers: onlookers can be *horrified* or *appalled*, but they would not be *outraged*. This suggests that we should not posit for *appalled* the component "I want to do something because of this."

Another difference between *outraged* and *appalled* can be illustrated with the following sentence:

They were appalled (*outraged) to see the suffering of the people in the wake of the floods.

In this sentence, *appalled* sounds natural, for it suggests that "something very bad happened to someone"; but *outraged* would sound odd—mainly, it seems, because it implies human action rather than a natural disaster. This shows that while *appalled* and *outraged* share the implication that "something very bad happened," *appalled* doesn't carry the further implication that it happened "because someone did something very bad."

Unlike *outraged*, *appalled* has also a reflective quality—as if one felt compelled to take note of and to reflect on terrible things that happen to people. The fact that *appalled* frequently co-occurs with the phrases "to see" or "to hear" (e.g., *I was appalled to see/hear* ...) highlights its "evidential" and compelling character. Wierzbicka (1999) accounted for this aspect with the component "I have to think now: Very bad things happen to people." On the other hand, *appalled* doesn't seem to imply a sudden discovery ("I know now"), characteristic of *outraged* (it is not perceived as a sudden experience, although the experiencer's attention is focused on some compelling evidence).

This brings us to the following explication:

4. *Appalled* (X was appalled) =
   (a) X felt something because X thought something
   (b)   sometimes a person thinks:
   (c)   "something very bad happened to someone
   (d)   I didn't think that something like this could happen
   (e)   I have to think now: very bad things happen to people"
   (f)   when this person thinks this this person feels something bad
   (g) X felt something like this
   (h) because X thought something like this

Constraints of space have permitted us to look briefly at only a handful of examples. The NSM literature, and especially Wierzbicka (1999), contains literally hundreds of detailed analyses of emotion terms in English and other languages. Nevertheless, I hope these few examples serve to make concrete the claim that the semantics of emotion terms can be "unraveled" in verbal paraphrases, providing that the analyses are solidly grounded in detailed linguistic observations and that the paraphrase technique is applied in a highly constrained and disciplined fashion. It is also pertinent to note that the method applies equally well to so-called "basic" emotion terms such as *sad*, which are often regarded—erroneously—as language universals, and to plainly language-specific terms such as *appalled*. Finally, notice that the approach is capable of delineating quite subtle differences in meaning between words like *sad* and *unhappy* that are normally regarded as synonyms. In general, it is doubtful whether there is any true or precise synonymy in the English emotion lexicon.

EXPLICATING EMOTION TERMS ACROSS LANGUAGES

The key project of linguistic semantics—that is, analyzing and stating meanings in a clear, intelligible, and testable way—becomes even more problematical once we cross a language barrier. Now, the other problems of semantic methodology are joined by the danger of ethnocentrism, that is, the danger of imposing a distorted interpretation on the facts of another language and culture due to unwittingly importing the assumptions, values, and the like, of the analyst's home culture. It is one of the strengths of the NSM method that it takes very seriously the danger of one particular form of ethnocentrism that often goes completely unnoticed by alternative methods. This is what we can call "terminological ethnocentrism": the imposing of a culture-specific analytical slant by the very choice of words employed as the metalanguage of description. This happens, for example, when the Yankunytjatjara words *pikaringanyi* and *mirpa ̲narinyi* are described as "kinds of *anger,*" when Yankunytjatjara itself contains no analytical category corresponding to English *anger.* Aside from the distorting effect, a description that is constructed in terms of language-specific words sacrifices an extremely valuable source of testing, because, being untranslatable into the language of the people concerned, it cannot be directly discussed with them or substituted into actual contexts of usage. Hence the value of the NSM program's commitment to a method of semantic description based entirely on semantic and lexical universals, that is, on meanings like *think*, *feel*, *want*, *good*, *bad*, *do* and so on, which evidence suggests have equivalent expressions in all languages.

To illustrate the kind of methods used and results obtained, we will in this section run through some examples of emotion semantics in languages other than English. As before, no particular theoretical weight attaches to this particular selection of examples, which are based on Goddard (1996) and Wierzbicka (1997a, 2001).

Malay "cemburu"

The Malay (Bahasa Melayu) emotion term *cemburu* is usually glossed as "jealous" or "envious." Typical textual examples do resemble the English terms. Consider:

Apa mungkin Sofi Aidura merajuk kerana ini? *Cemburu* kerana Sheila?

"Was it possible Sofi Aidura was sulking because of this? Was she *jealous* because of Sheila?"

... perli Kamalia tapi di hatinya adalah jugak perasaan *cemburu* sebab dia pun sedang mencari seseorang untuk memberi hatinya...

"...(she) teased Kamalia but in her heart there was a feeling of *envy* because she too was looking for someone to give her heart to..."

One may be *cemburu* over a lover or spouse, as in these examples, or one may be *cemburu* on account of someone else being better off either in material terms, for example, a neighbor getting a much better car than yours, or on account of someone's superior talent or other advantages, for example *Saya cemburu dengannya kerana kebolehannya* "I'm jealous of him on account of his abilities."

On the other hand, discussion with Malay speakers suggests some differences from both English *jealous* and *envious*. For one thing, one can hardly feel *cemburu* about someone one doesn't know personally, a difference (perhaps a slight one) from *envious*. Secondly, *cemburu* seems more active in orientation than either of its apparent English counterparts. That is, as a result of feeling *cemburu* one is inclined to do something—to berate one's spouse, to try harder at work or school, or whatever.

Some corroboration of this suggestion can be found in a large-scale study by Karl Heider of Indonesian emotion terms, as part of which he elicited typical antecedents and outcomes for *cemburu* and the related term *dengki*, which he glosses as "envy". Heider (1991) was able to contrast his findings with those of Davitz (1969) where comparable data was obtained about the outcomes of American English *jealousy*. The Indonesian terms, Heider found, "...have sharper, more malicious outcomes...with relatively little of the contemplative hurt and sadness, which are the most typical outcomes of at least one version of the American English emotions. One should not exaggerate the differences...yet the nuances are there (pp. 220-221).

From the point of view of an English speaker, the difficulty is finding an explication that is broad enough to cover the wide range of use of *cemburu*. It seems plain that the "envy scenarios" involve a comparative aspect. Roughly, one compares oneself with someone else who is "better off" in some respect, and feels bad as a result. But how to adjust this to accommodate the "jealous scenarios"? The apparent difficulty here arises from expectations based on the English concept of *jealousy*, which highlights personal loss and hurt. On asking Malaysian consultants about the kind of *cemburu* thoughts that might go through one's head if one's girlfriend, for instance, was flirting with someone else at a party, I was told: "Why does she want to go to another man, when I'm around?". In other words, even in this situation there is a comparison being made, and I come off the worse. A textual example is given below.

> Mengapa Jamal rapat dengan gadis yang bernama Sheila itu? Kenapa? Apa istimewanya gadis itu? Apa kurangnya aku?

> "Why is Jamal meeting with that girl Sheila? What's so special about that girl? What's wrong with me?"

We can therefore propose the following explication for *cemburu*.

5. *Cemburu* (X feels cemburu with Y) =
    (a)   X felt something because X thought something
    (b)   sometimes a person thinks about someone else:
    (c)   "this person is very good
    (d)   I am not like this
    (e)   I don't want this
    (f)   I want to do something because of this"
    (g)   when this person thinks this, this person feels something bad
    (h)   X felt something like this
    (i)   because X thought something like this

Polish "przykro"

The adverb *przykro* is one of the most important and salient of Polish emotion words. It is normally used in a dative construction, as in the following sentence, quoted in the Dictionary of the Polish Language (SJP):

> Przykro mi, że mimo woli obraziłem tę kobietę.
>
> *'I have a painful feeling (lit. (it is)* przykro *to-me) that I have unintentionally offended that woman.'*

The phrase *przykro mi* (literally, "it is painful to me") could be loosely translated into English in a number of ways, such as "I am sorry," "I feel bad," "I have an unpleasant feeling" or "I have a painful feeling." None of these would-be translations, however, captures the exact concept conveyed by the Polish phrase. As a first approximation, we could say that *przykro* conveys a painful feeling caused by a person and that to that extent it is similar to the English word *hurt*. But *hurt* implies an action done by someone else, whereas *przykro*—like English *sorry*—can also apply to some actions of the experiencer him or herself, as in the example above. Furthermore, *hurt* implies an unfulfilled expectation. One cannot feel *hurt* (in an emotional sense) by something that a stranger or an enemy does; one can only feel *hurt* as a result of something done by someone from whom one would expect "something good." Hence the oddity of sentences like: *?She was hurt by the dentist's/bus-driver's words.*

Przykro doesn't have such implications. In particular, something a dentist or a bus driver says can well cause one to feel *przykro*.

> Kiedy dentysta to powiedział, zrobiło mi się przykro.
>
> "When the dentist said this, I felt a painful feeling (I felt upset)' (lit. 'it became *przykro* to me)."

Przykro also differs from *sorry*, insofar as it necessarily refers to interpersonal relations. For example, I can be *sorry* when I hear that someone else lost their keys. *Przykro*, however, could not be used in such a context: It requires a human causer.

On the other hand, not all human actions can cause a feeling of *przykro*. Typically, the feeling is caused by somebody's words, as in the following example (with the noun *przykrość*):

Widocznie słowa moje sprawiły jej przykrość wielką, bo nerwowy, bolesny uśmiech wykrzywił chwilowo jej usta.

"One could see that my words caused her a great *przykrość*, because for a moment a painful, nervous smile twisted her mouth."

Nonverbal actions, and even omissions, can also cause the feeling of *przykro*—provided that the actions or omissions can be interpreted as conveying a particular message, a message of indifference, an absence of "good feelings" for someone else. For example, in the following sentence the person spoken of clearly interprets the speaker's behavior as sending such a message.

Już mi kilka razy mówił, że ja za dużo pracuję; widać więc, że i jemu przykro, że mnie tak mało widzi.

"He has already told me a few times that I work too much; it is clear, that he, too, feels upset (*przykro*) that he sees so little of me."

All the examples discussed are consistent with the following explication:

6. (*Przykro*) X-owi było przykro (X-DAT was painful-ADV) =
   (a) X felt something because X thought something
   (b)   sometimes a person thinks:
   (c)   "someone did something
   (d)   because of this, someone else could think:
   (e)   'when this person thinks about me, this person doesn't feel anything good'"
   (f)   when this person thinks this, this person feels something bad
   (g) X felt something like this
   (h) because X thought something like this

Most commonly, the causer, that is, the "someone" mentioned in the third line of the explication, is different from the experiencer, whereas the "someone else" in the fourth line is identical with that experiencer. It is also possible, however, for the experiencer to be identical with the causer. This is in fact one of the most striking features of the concept of *przykro*: that it encodes a bad feeling caused either by someone else's apparent emotional rejection of us (or indifference toward us), or by our own action which may appear to someone else to indicate our rejection of them (or indifference toward them). This brings us back to the first example, repeated here for convenience:

Przykro mi, że mimo woli obraziłem tę kobietę.

"I have a painful feeling that I have unintentionally offended that woman."

In this case, the speaker has done something, because of which someone else (the woman) could think: "when this person thinks about me this person doesn't feel anything good." In this case, the missing "good feelings" can be interpreted as respect—and as a result the woman feels offended. The speaker himself, thinking about what happened, feels *przykro*.

In an earlier study of the norms relating to emotion in Polish culture, Wierzbicka (1994d) argued that Polish culture places a great deal of emphasis on the expression of interpersonal "good feelings," noting that "the Polish cultural emphasis on warmth or affection must be distinguished from the Anglo emphasis on consideration and tact," and that "Polish culture encourages the showing of good feelings toward the addressee rather than attempts not to hurt or offend the other person" (p. 163). She also noted that "In Polish culture, behavior that shows feeling is seen as the norm, not as a departure from the norm" (p. 158), and that "perhaps no feeling is more valued and more expected in Polish discourse than a good feeling directed at the addressee" (p. 159). The culture-specific Polish concept of *przykro*, which focuses on painful effects of a perceived lack of expected interpersonal "good feelings," points to the same cultural values and expectations.

Japanese "amae"

According to Takeo Doi (1981), *amae* is "a thread that runs through all the various activities of Japanese society" (p. 169), which represents "the true essence of Japanese psychology" (p. 26) and is "a key concept for understanding Japanese personality structure" (p. 21). It is also a concept that provides "an important key to understanding the psychological differences between Japan and Western countries" (Doi, 1974, p. 310).

But what exactly is *amae*? Doi is convinced that there is no single word in English equivalent to it, a fact that the Japanese find hard to believe. Nonetheless, Doi and others have offered numerous clues that enable us to construct an English version of the concept of *amae*—not in a single word, of course, but in a semantic explication. Doi (1974) explained that "*amae* is the noun form of *amaeru*, an intransitive verb which means 'to depend and presume upon another's benevolence'" (p. 307). It indicates "helplessness and the desire to be loved" (Doi, 1981, p. 22). *Amaeru* can also be defined "by a combination of words such as 'wish to be loved' and 'dependency needs'" (1974, p. 309). The Japanese dictionary *Daigekan* defines *amae* as "to lean on a person's good will" (Doi, 1981, p. 72), or "to depend on another's affection" (1981, p. 167). Other dictionary glosses include "to act lovingly towards (as a much fondled child towards its parents)," "to presume upon," "to take advantage of," "to behave like a spoilt child," "be coquettish," "trespass-on," "take advantage of," "behave in a caressing manner towards a man," "to speak in a coquettish tone," "encroach on (one's kindness, good nature, etc.)," "presume on another's love," "coax," and so on.

Morsbach and Tyler (1986), who have analyzed passages from Japanese literature referring to *amae*, commented on the use of *amae* in these passages as follows: "As these fifteen examples illustrate, *amae* has a variety of meanings centering around passive dependency needs in hierarchical relationships" (p. 300). But the term "hierarchical relationship" is misleading. For example, it doesn't seem to fit the popular song (quoted by the authors) in which a female singer is asking her lover to permit her to *amaeru* to him:

> On the day we are finally one
> hug me, hug me,
> and you'll let me play baby, won't you?

Morsbach and Tyler pointed out "that at the time this song was popular there were no less than three pop tunes in which the word *amaeru* was used" (p. 296).

The most useful clue to the concept of *amae* is provided by the reference to the prototype on which this concept is based, a prototype that is not difficult to guess. "It is obvious that the psychological prototype of *amae* lies in the psychology of the infant in its relationship to its mother"; not a newborn infant, but an infant who has already realized "that its mother exists independently of itself ...[A]s its mind develops it gradually realizes that itself and its mother are independent existences, and comes to feel the mother as something indispensable to itself, it is the craving for close contact thus developed that constitutes, one might say, *amae*" (Doi, 1981, p. 74). This is the prototype. But according to Doi and others, in Japan the kind of relationship based on this prototype provides a model of human relationships in general, especially (though not exclusively) when one person is senior to another.

> He may be your father or your older brother or sister ... But he may just as well be your section head at the office, the leader of your local political faction, or simply a fellow struggler down life's byways who happened to be one or two years ahead of you at school or the university. The *amae* syndrome is pervasive in Japanese life. (Gibney, 1975, p. 119)

Given all the complexity and versatility of the *amae* concept some have doubted that a unitary definition of it can be given at all. However, although it is evident that *amae* has a fairly complex semantic structure, a comprehensive explication is nevertheless possible.

(7) *Amae* =
    (a) X felt something because X thought something
    (b)   sometimes a person thinks something like this:
    (c)   "when Y thinks about me, Y feels something good
    (d)   Y wants to do good things for me
    (e)   Y can do good things for me
    (f)   when I am with Y nothing bad can happen to me
    (g)   I don't have to do anything because of this"
    (h)   when this person thinks this, this person feels something good
    (i) X felt something like this
    (j) because X thought something like this

The phrasing of the first component reflects the need for conscious awareness. The presumption of a "special relationship" is reflected in the component "when Y thinks about me, Y feels something good." The implication of self-indulgence is rooted in the emotional security of someone who knows that he or she is loved: "it is an emotion that takes the other person's love for granted" (Doi, 1981, p. 168). This is accounted for by the combination of components: 'Y wants to do good things for me," "Y can do good things for me," and "when I am with Y nothing bad can happen to me." The component "I don't have to do anything because of this" reflects the "passive" attitude of an *amae* junior, who does not need to earn the mother-figure's goodwill and protection by any special actions.

A great deal has been written on the reasons for the prominence of *amae* in Japanese society. According to Doi himself (1981), this is linked with an "affirmative attitude toward the spirit of dependence on the part of the Japanese" (p. 16). Numerous observations in the literature point in the same direction. For example, Murase (1984) notes that "Unlike Westerners, Japanese children are not encouraged from an early age to emphasize individual independence or autonomy. They are brought up in a more or less 'interdependent' or *amae* culture ..." (p. 319). Clearly, perceptions of this kind are highly consistent with the prominence of the feelings of trustful dependence elucidated by Doi and others.

In this section, we have seen three examples of salient culture-specific emotion words from non-English languages. The existence of these words does not, of course, mean that there is a unbridgeable gulf between the emotional worlds of people who speak English, Malay, Polish, and Japanese. To some extent at least, emotional experience is probably similar across these four cultures. However, what the differences in lexicon mean is that people from these four "linguistic cultures" speak about and think about their emotional experiences very differently. The same situation may be conceptualized in different terms. For example, a situation that a Polish speaker might conceptualize in terms of the word *przykro* would be likely to be conceptualized in English in a different way—for example, as a situation linked with the experience of being *hurt, offended, sorry,* or *feeling bad.*

The general point is that each language provides its speakers with a different set of interpretive categories and encourages them to conceptualize their emotional experience in terms of a different set of scenarios. Although we have not been able to pursue this point at any length here, it can be argued convincingly that the set of scenarios linked with frequently used language-specific lexical categories (so-called "cultural key words") points to a culture's central concerns and values (cf. Wierzbicka, 1997b).

CULTURAL SCRIPTS ABOUT EXPRESSING FEELINGS

So far in this chapter we have stayed within the realm of lexical semantics, that is, we have been concerned with explicating the language-specific meanings of individual words and, to a very limited extent, with correlating these meanings with the values and priorities of the culture. In this section, we move away from lexical semantics into the realm of "cultural pragmatics." This concept needs a few words of introduction. The basic insight, long recognized by anthropological linguists and ethnographers of communication, is that different speech communities have different "ways of speaking," not just in the narrowly linguistic sense but also in the norms or conventions of linguistic interaction. For example, in some parts of the world it is quite normal for conversations to be loud, animated, and bristling with disagreement, whereas in others people avoid contention, speak in well-considered phrases, and guard against exposure of their inner selves. In some places, silence is felt to be awkward, while in others silence is welcomed. In some societies it is considered very bad to speak when another person is talking, while in others this is an expected part of a co-conversationalist's work. Needless to say, some variations of this kind concern the expression of emotions.

How are we to describe these different "local" conventions of discourse? Linguists, anthropologists, and other scholars have employed a variety of frameworks. Most assume that within any particular speech community there are certain tacitly shared understandings about how it is appropriate to speak in particular, culturally construed situations. The "cultural scripts" method, described in this section, is an approach to stating these inferred shared understandings (Wierzbicka, 1994b, 1994c, 1996b, 1997a).

The basic idea is a simple one. Instead of employing descriptive labels such as "direct" and "indirect," "polite," "formal," and so on, to characterize the speech patterns of different cultural and social settings, we employ the independently motivated metalanguage of universal semantic primes. As we will see, this means that the particular cultural norms can be spelled out in much greater detail than is possible with global labels such as "direct," "formal," and so on, which is a virtue in itself. But another advantage is that cultural scripts are formulated in terms that are recognizable and intelligible to the people concerned. In a sense, a cultural script is a hypothesis about cultural assumptions, that is, about something that is internalized by speakers as they are socialized into a particular lifestyle, value system, and set of social practices.

It is important to stress that cultural scripts are not intended to provide an account of real life social interactions, but rather as descriptions of commonly held assumptions about what "people think" about social interaction. Social actors bring these assumptions with them into everyday interactions. As such, cultural scripts influence the form taken by particular verbal encounters. However, they do not in any sense determine individual interactions. Individuals can and do vary in their speech behavior. The claim of the cultural scripts approach is merely that the scripts form a kind of interpretative background against which individuals position their own acts and those of others.

In connection with emotions, perhaps the most relevant cultural scripts are those that relate directly to emotional expressivity. We now look at some examples from the same languages as used earlier, examples from Wierzbicka (1997a, 1999) and Goddard (1997a).

Polish vs. American Cultural Scripts on "Saying What You Feel"

Broadly speaking, Anglo-American culture values and encourages the display of "good feelings" that one may not necessarily feel, and the suppression of "bad feelings" whose display may be seen as serving no useful purpose and either damaging to our "image" or unpleasant for other people. Polish culture, in contrast, values saying and showing what one really feels, even at the cost of causing unpleasantness or hurt to others. In particular, the two cultures have different norms and expectations concerning smiling. An American woman married to a Pole and living in Warsaw (Klos Sokol, 1997) wrote that "Americans smile more in situations where Poles tend not to"; Poles don't "*initiate* an exchange of smiles in a quick or anonymous interaction"; in Poland "you may see faces that might look *really* grumpy" (p. 119). She commented further:

> In everyday life, the approach to fleeting interactions in Poland is often take-me-seriously. Rather than the cursory smile, surface courtesy means a slight nod of the head. And some Poles may not feel like masking their everyday preoccupations. From this perspective, the smile would be fake. In American culture, you don't advertise your daily headaches; it's bad form; so you turn up the corners of the mouth—or at least try—according to the Smile Code. (p. 118)

The tacit assumption behind what Klos Sokol calls the American "Smile Code" can be represented in the form of the following cultural script:

8. An Anglo-American cultural script
    people think:
    when I say something to other people
    it is good if these people think that I feel something good

Of course, it goes without saying that not all Americans live by this script, but they are surely all familiar with it. The component "people think:" that opens the scripts reflects the fact that even people who personally don't identify with the content of the script are nonetheless familiar with it. They, too, belong to the community that shares familiarity with this script (and with other, related, cultural scripts.) In Polish culture, however, there is no similar (generally recognizable) tacit assumption.

The evidence just adduced is of course purely anecdotal and subjective. There is, also, however, objective linguistic evidence that points in the same direction. This evidence includes, in particular, the strongly negative connotations of Polish words like *fałszywy* ("false") and *sztuczny* ("artificial"), used for condemnation of "put-on" smiles and other forms of non-spontaneous displays of "good feelings". The collocation *fałszywy uśmiech,* "a false smile," is particularly common, but others include *fałszywe pocałunki* "false kisses," *fałszywy uśmiech* "a false smile," and *sztuczny uśmiech* "an artificial smile." Such expressions all imply that someone is displaying good feelings towards another person that in fact are not felt, and that "of course" it is very bad to do so. This can be represented in the form of the cultural script in (9).

9. A Polish cultural script:
    people think:
    it is bad
    when a person wants other people to think that this person feels something
    if this person doesn't feel this

One implication of this script is that a person's face should reflect his or her feelings, whatever they may be. The cultural assumption is summed by a quote from the writer Niemcewicz, adduced by SJP:

> Uczucia, które sztucznie udajemy, serca nasze oziębiają.

> 'Feelings that we artificially display, make our hearts cold.'

Both sociological analysis and linguistic evidence concur with cross-cultural experience that Anglo-American attitudes value "displays of good feelings" in general. What applies to smiles applies also to what may be called "cheerful speech routines." In English, there are many common speech routines that manifestly reflect a cultural premise to the effect that it is good to feel good—and to be seen as someone who feels good. For example, the common *How are you? — I am fine* routine implies an expectation that good feelings will be expressed, and if need be, "artificially" displayed. Anglo-American culture appears to have gone further in the direction of positive scripts than the Anglo-British or Anglo-Australian varieties have, and has apparently developed some emotional scripts of its own, two of which could be called "the enthusiasm script" and the "cheerfulness script." To quote an American witness (Klos Sokol, 1997, p. 176) again:

> Wow! Great! How nice! That's fantastic! I had a terrific time! It was wonderful! Have a nice day! Americans. So damned cheerful.

One linguistic reflection of this attitude is the ubiquitous presence of the word *great* in American discourse (cf. Wolfson, 1983, p. 93), both as a modifier (especially of the verb *to look*) and as a "response particle":

> You look great!
> Your X (hair, garden, apartment, etc.) looks great!
> It's great! That's great! Great!

Speech routines of this kind suggest a cultural script that can be formulated along the following lines:

10. Anglo-American cultural script:
    people think:
    it is good to say often something like this:
        "I feel something very good"

It is interesting to note in this connection that, for example, Sommers' (1984) cultural study of attitudes toward emotions showed that Americans place an exceptional emphasis on "enthusiasm" and value it far more highly than do the other cultural groups with which they were compared (Greeks, West Indians, Chinese). In a similar vein, Renwick (1980, p. 28) contrasted the "Australian art of deadpan understatement" with the American penchant for "exaggeration and overstatement."

Polish doesn't have speech routines corresponding to *How are you? — I'm fine*, or to the ubiquitous American *Great!*. This is consistent with the subjective evidence such as that reported by Klos Sokol (1997):

> One Pole said, "My first impression was how happy Americans must be." But like many Poles she cracked the code: "Poles have different expectations. Something 'fantastic' for Americans would not be 'fantastic' in my way of thinking." Another Pole says, "When Americans say it was great, I know it was good. When they say it was good, I know it was okay. When they say it was okay, I know it was bad." (p. 176)

The central importance of positive feelings in American culture is also reflected in the key role that the adjective *happy* plays in American discourse, an adjective that is widely used as a yardstick for measuring people's psychological well-being as well as their social adjustment. The crucial role of this adjective in

American life has often been commented on by newcomers. For example, Stanisław Barańczak (1990), professor of Polish literature at Harvard University, wrote:

> Take the word *happy*, perhaps one of the most frequently used words in Basic American. It's easy to open an English-Polish or English-Russian dictionary and find an equivalent adjective. In fact, however, it will not be equivalent. The Polish word for *happy* (and I believe this also holds for other Slavic languages) has much more restricted meaning; it is generally reserved for rare states of profound bliss, or total satisfaction with serious things such as love, family, the meaning of life, and so on. Accordingly, it is not used as often as *happy* is in American common parlance... (p. 13)

The pressure on people to be *happy* can only be compared with the pressure to smile: By being *happy*, one projects a positive image of oneself (as a successful person).

To be *happy* is to feel something good for personal reasons—an ideal quite consistent with the general orientation of "a culture dominated by expressive and utilitarian individualism" (Bellah, Madsen, Sullivan, Swidler, & Tipton, 1985, p. 115). The fact that *happy* is an adjective, whereas its closest counterparts in other European languages are verbs (e.g., *sich freuen* in German, *se rejouir* in French, or *cieszyćsię* in Polish) is also significant, because these verbs indicate a temporary occurrence (as the archaic verb *rejoice* does in English), whereas the adjective *happy* is compatible with a long-term state (the expected norm). As Barańczak points out, people can be expected to be *happy* most of the time, but can not be expected to *rejoice* most of the time.

To quote another cross-cultural observer, Eva Hoffman (1989), who migrated from Poland to America as a teenager wrote:

> If all neurosis is a form of repression, then surely, the denial of suffering, and of helplessness, is also a form of neurosis. Surely, all our attempts to escape sorrow twist themselves into the specific, acrid pain of self-suppression. And if that is so, then a culture that insists on cheerfulness and staying in control is a culture that—in one of those ironies that prevails in the unruly realm of the inner life—propagates its own kind of pain. (p. 271)

Hoffman's assessments of the psychological costs of obligatory cheerfulness may or may not be correct, but few commentators would disagree with the basic idea that something like cheerfulness is encouraged by American culture. More precisely, the norm in question can be represented as follows:

11. An Anglo-American cultural script:
     people think:
     it is good to think often that something good will happen
     it is good to often feel something good because of this
     it is good if other people can see this

What can one do to comply with the above norm? One can, of course, smile—and, as we have seen, American culture is one of those cultures that value and encourage the "social smile."

Malay Cultural Scripts for Emotional Caution and Sensitivity

In general, it can be said that Malay culture discourages people from verbally expressing how they feel, the ideal demeanor being one of good-natured calm, in keeping with the cultural ideal of *senang hati* (Goddard, 1997b). From a Western perspective, in most social relationships people cultivate "unemotional presentation" (Banks, 1983, p. 88). On the other hand, everyone is expected to be sensitive to other people's facial expressions and actions or be deemed *bodoh* "thick, stupid."

The script in (12) effectively discourages verbal explicitness about one's feelings, while expressing confidence in the effectiveness of non-verbal signals.

12. A Malay cultural script:
    people think:
    when I feel something
    it is not good to say something like this to another person:
        "I feel like this"
    if the other person can see me, they will know how I feel

The use of "meaningful looks" (*pandangan bermakna*) is a favored nonverbal strategy. For instance, the verb *tenung* (cf. *bertenung* "to divine") depicts a kind of glare used to convey irritation with someone else's behavior, for example, a child misbehaving or someone in the room clicking a pen in an irritating way. Widening the eyes *mata terbeliak* (literally "bulging eyes") conveys disapproval. Lowering the eyes and deliberately turning the head away (*jeling*) without speaking can convey that one is "fed up" with someone. Pressing the lips together and protruding them slightly (*menjuihkan bibir*) conveys annoyance.

Nonverbal expression is critical to the closest Malay counterpart of English "angry," namely *marah* "offended, angry." This is associated not with scenes of "angry words" as sanctioned by Anglo cultural scripts of free self-expression, but by the sullen brooding performance known as *merajuk*.

Malay reluctance to verbalize about feelings can also be seen if we consider the situation in which I realize that I have done something bad to someone else. It would be difficult to say outright the equivalent of "I'm sorry, I was wrong." The preferred strategy is to be extra nice to the person in question, who will understand. If I must say something, it should be vague and there should be no direct reference to your feelings or to mine. One could say something like this:

Kalau semalam aku ada terbuat/tercakap yang kasar aku minta maaf yelah.

"If yesterday I did/said something rough I ask for pardon, yes."

Needless to say, there is no exact Malay equivalent to the English speech-act verb "apologize." The speech act above would be classed as *pujuk*, a word whose range also takes in "comfort, console," that is to say, it is focused on making the other person feel better rather than on displaying one's own regrets. The expression *minta maaf*, "ask pardon" sounds somewhat formal and does not to correspond to English "saying sorry."

Of course, both these examples involve the speaker's "bad feelings," and in scenarios in which the addressee too could conceivably feel something bad. There is abundant evidence, however, that Malay cultural ideology disfavors verbalizing about "good feelings" also. This is particularly clear in formal situations. At a Malay wedding, for instance, it is bad form for either the bride or the groom to smile. They are supposed to maintain a composed, calm expression. Karim (1990a) cites Wilder (1982) approvingly:

> During marriage, as in other contexts of Malay social relations, constant emphasis is placed on the maintenance of personal and social equilibrium and restraint. In Malay social relations in general, as in marriage in particular, a keenly-felt balance and reserve operate to counter the public display of affection, or hostility, or practically any deeper emotion. (p. 74)

The same applies during courtship. It is considered bad for the young people to give explicit signs of their affection for one another. As Karim (1990b) said, "during courtship, a person regardless of sex, has to be careful to conceal his or her feelings in public" (p. 29). To do otherwise would invite *malu* "shame." When young people betray observable signs of infatuation or love, these are seen as "disorders" (*gila*, also "craziness").

Interestingly, Karim (1990b) highlighted the role of the *pantun* in controlling the display of emotion during courtship. The *pantun* is a Malay poetic form, consisting of two rhyming couplets in which an emotion or mood is implied or evoked. Fauconnier (1990/1931) places great emphasis on their role in traditional rural Malay life:

> It is the play on words, the equivocations, the tenuous allusions, that constitute their special charm for the Malays. ... They all know a large number of pantuns and are constantly inventing new ones. Their conversation is full of these poetic insubstantial images. (p. 82)

One can easily appreciate how an abundance of pantuns, and of other evocative *peribahasa* "sayings," would be a very serviceable resource for alluding to potentially sensitive matters. In traditional-style courtship they play a special role, providing the young man and woman with a compact written medium in which their feelings may be expressed with acceptable restraint. Karim (1990b) described how a series of such veiled messages will often be passed between the pair by a go-between. She commented, "Like Japanese haiku, the brevity of verse attempts a controlled elegance over emotion. It is in this sense a mode of communication which guides passion into acceptable poise and restraint" (p. 32). The comparison with the haiku is highly suggestive, Japanese culture being widely known as one which disfavors displays of emotion.

Two situations where European norms would lead one to expect "good feelings" to be expressed verbally are accepting a gift and in response to a compliment, but Malay cultural norms differ in both cases. On accepting a gift (after the appropriate refusals) one would never say anything like "Oh wow, how great!" in the Anglo-American style. It would be good form instead to look a little to one side, perhaps smiling slightly, and to softly say *terima kasih* (*terima* "receive," *kasih* "care"), which though usually translated as "thank you" is better understood as an acknowledgement of the other person's kindness.

If someone says something nice about you, for instance, that your shirt or dress looks nice or looks new, one would deflect the remark along the following lines:

Takde lah. Ini kan yang Cliff beli tahun lalu.

"This is nothing, I bought it years ago."

Takde lah. Biasa aja.

"This is nothing, it's just ordinary."

If it is something more personal, such as that you are a good cook, or that you are skilled at your work, or (less common) that you are pretty, the right thing to do is not to say anything, and not to look directly at the other person either. One looks downward and a little to the side, perhaps smiling slightly (*tersenyum sedikit*). Of course, these responses could (and should) be spelled out in more specific scripts. My point here is just that the favored mode of nonverbal response is consistent with a general restriction on directly verbalizing one's feelings.

Cultural commentators invariably mention that Malay culture greatly values the capacity of a person to be "sensitive," "considerate," and "understanding" of others, and therefore to always speak with care lest the other person has his or her feelings hurt (*tersinggung*). For example, Wilson's (1967) list of Malay values includes the following: "showing consideration and concern, anticipating the other ... and, above all, being sensitive to the other person" (pp. 131-132). Rogers (1993) stressed "the great emphasis placed on harmonious personal relations in Malay culture" (p. 30, 42). Many traditional sayings enjoin people to watch over (*jaga*) other people's feelings (*jaga hati orang*) or to "look after feelings" (*memilihara perasaan*).

Evidence like this suggests a cultural script like the following:

13. A Malay cultural script:
   people think:
   before I say something to someone, it is good to think:
      I don't want this person to feel something bad because of this

What could make a person "feel something bad"? In the Malay sociocultural context, one of the most powerful of bad feelings is that of *malu*. Though often glossed as "shame" this word has a much broader range of use than the English word "shame", and it also lacks the negative connotations. If anything, Malays regard a sense of *malu* as a positive social good. I have argued elsewhere (Goddard, 1996) that the lexical meaning of *malu* can be characterized, in large part, as an unpleasant and unwanted feeling due to the thought that other people are thinking and saying bad things about one.

The cultural importance of *malu* is confirmed by the fact that there is a whole cluster of associated concepts, such as *maruah* "dignity," *nama* "reputation," and *air muka* "face" (literally water face), concerned with one's standing in the eyes of others; in others words, about what other people think about one. Observers of Malay culture have often remarked on the salience of such concepts, both in interpersonal interaction and at the level of public discourse. For example:

> The social value system is predicated on the dignity of the individual and ideally all social behavior is regulated in such a way as to preserve one's own amour propre and to avoid disturbing the same feelings of dignity and self-esteem in others (Vreeland et al. 1977, p. 117).

The cultural history of *nama* as a "key concept" in the development of modern Malaysian politics has been traced in persuasive detail by Milner (1982, 1995).

Representative of various sayings that stress the importance of *nama* and related concepts is the following:

> Gajah mati meninggalkan tulang, harimau mati meninggalkan belang, manusia mati meninggalkan nama.

> "When elephants die they leave behind bones, when tigers die they leave behind stripes, when humans die they leave behind nama."

The upshot of all this, from the point of view of emotional expression, is that one is well-advised to be careful about saying anything publicly that could lower another person's *maruah* or *nama*, and especially so if that person is in a position to retaliate (or has supporters who are likely to do so).

Japanese Cultural Scripts About Causing "Bad Feelings" in Others

In the literature on Japanese culture and society, it is often said that in Japan it is important to apologize very frequently and in a broad range of situations. The experience of Western students of Japanese is consistent with such statements. As Coulmas (1981) reported,

> [A] Western student who has been taught Japanese experiences the extensive usage of apology expressions as a striking feature of everyday communication when he first comes to Japan. Correspondingly, "Among Japanese students of English, German, or other European languages, it is a common mistake to make apologies where no such acts are expected or anticipated in the respective speech community. (p. 81)

The Japanese psychiatrist Takeo Doi (1981) recalled in this connection an observation made by the Christian missionary Father Henvers about "the magical power of apology in Japan," and he commented: "It is particularly noteworthy that a Christian missionary, who came to Japan to preach forgiveness of sin, should have been so impressed by the realization that among Japanese a heartfelt apology leads easily to reconciliation" (p. 50). To illustrate this point, Doi recounted the experience of an American psychiatrist in Japan, who through some oversight in carrying out immigration formalities, "found himself hauled over the coals by an official of the Immigration Bureau." However often he explained that it was not really his fault, the official would not be appeased, until, at the end of his tether, he said "I'm sorry ..." as a prelude to a further argument, whereupon the official's expression suddenly changed and he dismissed the matter without further ado." Doi concludes his discussion with a characteristic comment that "people in the West (...) are generally speaking reluctant to apologize." (p. 51).

But observations such as those made by Coulmas and Doi, though revealing, are not specific enough. To begin with, the concept of "apology" itself is culture-bound and is therefore inappropriate as a descriptive and analytical tool in the cross-cultural field. The words *apology* and *apologize*, which are elements of the English set of speech act terms, include in their meaning the component "I did something bad (to you)." But as Doi's little anecdote illustrates, the so-called "Japanese apology" does not presuppose such a component. It is misleading and confusing, therefore, to call it "apology" in the first place.

Furthermore, those who talk of the extensive usage of apologies in Japan (as compared with the West) create an impression that the difference is quantitative, not qualitative. This is misleading and inaccurate: in fact, the difference lies not in the frequency of use of the same speech act, but in the use of qualitatively different speech acts, and the use of these different speech acts is linked with qualitatively different cultural norms. Norms of this kind can be usefully illustrated with schematic scenarios, such as those offered in Kataoka and Kusumoto's (1991) book *Japanese Cultural Encounters and How to Handle Them.*.

Tom rented a car one weekend. It was his first time driving a car in Japan, but he had been an excellent driver in the United States.

On his way to a friend's house, however, he had an accident. A young child about four years old ran into the street from an alley just as Tom was driving by. Tom was driving under the speed limit and he was watching the road carefully, so he stepped on the brakes immediately. However, the car did brush against the child, causing him to fall down. Tom immediately stopped the car and asked a passerby to call the police and an ambulance.

Fortunately, the child's injuries were minor. The police did not give Tom a ticket, and he was told that he was not at fault at all, thanks to some witnesses' reports. He felt sorry for the child but decided that there was nothing more he could do, so he tried to forget about the accident. However, after several days, Tom heard from the policeman that the child's parents were extremely upset about Tom's response to the incident. (p. 2)

Kataoka invited the reader to consider four alternative answers to the question "Why were the child's parents upset?" The following answer is then indicated as the correct one: "They were angry because Tom did not apologize to them, nor did he visit the child at the hospital, even though he was not at fault. Tom should have done these things to show his sincerity." Kataoka commented further: "In Japan, one is expected to apologize and visit the victim of an accident, even if one is not at fault, to show his or her sincerity. In fact, one is expected to apologize whenever the other party involved suffers in any way, materially or emotionally. In many court cases, perpetrators get a lighter sentence when it is clear that they regret their actions, as reflected in their apology." (p.64).

The cultural norm reflected in Kataoka's story and explanatory comments can be represented in the form of the following cultural script:

14. A Japanese cultural script:
    people think:
    if something bad happens to someone because I did something
    I have to say something like this to this person:
        "I feel something bad"
    I have to do something because of this

The cultural rule in question was clearly illustrated by the sudden resignation (April 8, 1994) of the Japanese Prime Minister Mosihiro Hosokawa. According to newspaper reports, Mr. Hosokawa said that "the scandal over his financial dealings was 'extremely regrettable' because it had prevented the Parliament from passing the budget and hindered his reform plans" (*The Australian*, p. 1). "Mr. Hosokawa said there was nothing wrong with the two loans he accepted during the 1980s, but he felt morally responsible for the parliamentary impasse" (p. 12). Thus, Mr. Hosokawa denied that he had done anything bad, but he admitted that something bad (a parliamentary impasse) happened because he had done something (accepted two loans). This admission made it necessary for Mr. Hosokawa to say, publicly, that he felt something bad because of what had happened, and this, in turn, made it necessary for him to do something (resign), to show that he really did feel something bad (that is, "to prove his sincerity"). Thus, the cultural scenario enacted by the Prime Minister corresponds exactly to the one which should have been enacted in Kataoka's story.

The importance of paying attention to other people's "bad feelings" that we may have caused is reflected in a number of other cultural rules that can only be mentioned briefly here. One often commented on has to do with the so-called "blurring of apologies and thanks" in Japanese culture. In Anglo cultures, there is a basic rule that, roughly speaking, requires people to respond to favors in a positive way, as in (15a). This is in direct contrast with situations when we have to apologize to other people, which can be characterized, roughly, as in (15b).

15a. An Anglo cultural script:
  people think:
  when someone does something good for me
  I have to say something like this to the person:
    "I feel something good because of this"

15b. An Anglo cultural script:
  people think:
  when I do something bad to someone
  I have to say something like this to this person:
    "I feel something bad because of this"

But in Japanese culture, there is no similar contrast between the two types of situations, and in both a negative response is appropriate:

16a. A Japanese cultural script:
  people think:
  when I do something bad to someone
  I have to say something like this to this person:
    "I feel something bad because of this"

16b. A Japanese cultural script:
  people think:
  when someone does something good for me
  it is good to say something like this to this person:
    "I feel something bad because of this"

The Japanese script that links reception of favors with the need to express "bad feelings," is puzzling to Westerners, but from the point of view of Japanese cultural logic, this script makes perfect sense because it reflects the speakers' awareness of the trouble that they may have caused. As Coulmas (1981) wrote, "The Japanese conception of gifts and favors focuses on the trouble they have caused the benefactor rather than the aspects which are pleasing to the recipient" (p. 83). The same script explains also why, as Coulmas (1981) pointed out, Japanese dinner guests on leaving would say something like "I have intruded on you" or "Disturbances have (been) done to you" rather than "Thank you so much for the wonderful evening" (p. 83).

CONCLUDING REMARKS

What I have tried to do in this chapter is to sketch out the integrated and meaning-based approach to the study of emotions that has been pioneered by Anna Wierzbicka. It seeks to bring together the study of the emotion lexicon of different languages with the study of different "cultural scripts" that are one factor (among others, of course) influencing the expression of emotions in discourse. More than this, it also aims (though we have not touched on these matters in this chapter) to take in the encoding of emotional meanings by means of other linguistic devices, such as exclamations and specialized grammatical constructions, and even the encoding of emotional meanings in facial expressions and kinaesthetics (Hasada, 1996b; Wierzbicka 1995a, 1995b). Because the natural semantic metalanguage is based on simple, universally available meanings, it provides a tool that enables us to undertake this very broad range of investigations across languages and cultures, while minimizing the risk of ethnocentrism creeping into the very terms of description.

REFERENCES

Aebischer, V. & Wallbott, H. G. (1986). Measuring emotional experiences: questionnaire design and procedure, and the nature of the sample. In K. R. Scherer, H. G. Wallbott, & A. B. Summerfield (Eds.), *Experiencing emotion. A cross-cultural study (pp. 28-38).* Cambridge: Cambridge University Press.

Ameka, F. (1990). The grammatical packaging of experiencers in Ewe: A study in the semantics of syntax. *Australian Journal of Linguistics, 10,* 139-181.

Banks, D. (1983). *Malay kinship.* Philadelphia: Institute for the Study of Human Issues.

Barańczak, S. (1990). *Breathing Under Water and Other East European Essays.* Cambridge, MA: Harvard University Press.

Bellah, R. N., Madsen, R., Sullivan, W. M., Swidler, A. & Tipton, S. M. (Eds.). (1985). *Habits of the heart: Individualism and commitment in American life.* Berkeley: University of California Press.

Bruner, J. (1990). *Acts of meaning.* Cambridge, MA: Harvard University Press.

Chappell, H. (1986). The passive of bodily effect in Chinese. *Studies in Language, 10,* 271-296.

Coulmas, F. (Ed.). (1981). Poison to your soul: Thanks and apologies contrastively viewed. In F. Coulmas (Ed.) *Conversational routine* (pp. 69-91). The Hague: Mouton.

Davitz, J. R. (1969). *The language of emotion.* New York: Academic Press.

Doi, T. (1974). Amae: A key concept for understanding Japanese personality structure. In T. S. Lebra & W. P. Lebra (Eds.), *Japanese culture and behavior* (pp. 145-154). Honolulu: University of Hawaii Press.

Doi, T. (1981). *The anatomy of dependence.* Tokyo: Kodansha.

Ekman, P. (1992). An argument for basic emotions. *Cognition and Emotion, 6,* 169-200.

Ekman, P. (1994). All emotions are basic. In P. Ekman & R. J. Davidson (Eds.), *The nature of emotion: Fundamental questions* (pp. 15-19). New York/Oxford: Oxford University Press.

Ekman, P. & Davidson, R. J. (Eds.). 1994. *The nature of emotion: Fundamental questions.* New York/Oxford: Oxford University Press.

Fauconnier, H. (1990/1931). *The soul of Malaya.* (E Sutton, Trans.) Singapore: Oxford University Press.

Fehr, B. & Russell, J. (1984). Concept of emotion viewed from a prototype perspective. *Journal of Experimental Psychology: General, 113,* 464-486.

Gibney, F. (1975). *Japan: The fragile superpower.* New York: Norton.

Goddard, C. (1991). Anger in the Western Desert: A case study in the cross-cultural semantics of emotion. *Man, 26,* 265-279.

Goddard, C. (1995). "Cognitive mapping" or "verbal explication"?: Understanding love on the Malay Archipelago. *Semiotica, 106,* 323-354.

Goddard, C. (1996). The "social emotions" of Malay (Bahasa Melayu). *Ethos, 24,* 426-464.

Goddard, C. (1997a). Contrastive semantics and cultural psychology: 'Surprise' in Malay and English. *Culture & Psychology, 3,* 153-181.

Goddard, C. (1997b). Cultural values and 'cultural scripts' of Malay (Bahasa Melayu). *Journal of Pragmatics, 27,* 183-201.

Goddard, C. (1998). *Semantic analysis. A practical introduction.* Oxford: Oxford University Press.

Goddard, C. & Wierzbicka, A. (Eds.) 1994. *Semantic and lexical universals—Theory and empirical findings.* Amsterdam/Philadelphia: John Benjamins.

Goddard, C. & Wierzbicka, A. (Eds.). (in press). *Meaning and universal grammar.* Amsterdam: John Benjamins.

Harkins, J. (1990). Shame and shyness in the Aboriginal classroom: A case for "practical semantics." *Australian Journal of Linguistics* (Special Issue on Emotions), *10,* 293-306.

Harkins, J. (1996). Linguistic and cultural differences in concepts of shame. In D. Parker, R. Dalziell, & I. Wright (Eds.), *Shame and the modern self* (pp. 84-96). Melbourne: Australian Scholarly Publishing.

Harkins, J., & Wierzbicka, A. (Eds.) (in press). *Emotions in cross-linguistic perspective.* Berlin: Mouton de Gruyter.

Harré, R. (Ed.). (1986). *The social construction of emotions.* Oxford: Basil Blackwell.

Harris, P. (1989). *Children and emotion: The development of psychological understanding.* Oxford: Basil Blackwell.

Hasada, R. (1994). *The semantic aspect of onomatopoeia: Focusing on Japanese psychomimes.* Unpublished masters thesis. The Australian National University.

Hasada, R. (1996). Some aspects of Japanese cultural ethos embedded in nonverbal communicative behaviour. In F. Poyatos (Ed.), *Nonverbal communication in translation* (pp. 83-103). Amsterdam: John Benjamins.

Hasada, R. (1998). Sound symbolic words in Japanese. In A. Athanasiadou & E. Tabakowska (Eds), *Speaking of emotions: Conceptualisation and expression* (pp. 83-98). Berlin: De Gruyter.

Heider, K. (1991). *Landscapes of emotion. Mapping three cultures of emotion in Indonesia.* Cambridge: Cambridge University Press.

Hoffman, E. (1989). *Lost in Translation: A life in a new language.* London: Minerva.

Karim, W. J. (1990a). Introduction: emotions in perspective. In W. J. Karim (Ed.) *Emotions of culture. A Malay perspective* (pp. 1-20). Singapore: Oxford University Press.

Karim, W. J. (1990b). Prelude to madness: The Language of emotion in courtship and early marriage. In W. J. Karim (Ed.) *Emotions of culture. A Malay perspective* (pp. 21-63). Singapore: Oxford University Press.

Kataoka, H. C. & Kusumoto, T. (1991). *Japanese cultural encounters and how to handle them.* Chicago: Passport Books.

Kessler, C. S. (1992). Archaism and modernity: Contemporary Malay political culture. In J. S. Kahn & F. Loh Kok Wah (Eds.), *Fragmented vision* (pp. 133-157). Sydney: Allen and Unwin.

Klos Sokol, L. (1997). *Shortcuts to Poland.* Warsawa: IPS Wydawniclwo.

Lazarus, R. S. (1991). *Emotion and adaptation.* New York: Oxford University Press.

Lazarus, R. S. (1995). Vexing research problems inherent in cognitive-mediational theories of emotion and some solutions. *Psychological Inquiry, 6,* 183–196.

Levy, R. (1973). *Tahitians: Mind and experience in the Society Islands.* Chicago: Chicago University Press.

Lutz, C. (1988). *Unnatural emotions: Everyday sentiments on a Micronesian atoll and their challenge to Western theory.* Chicago: Chicago University Press.

Milner, A. (1982). *Kerajaan: Malay political culture on the eve of colonial rule.* Tuscon: University of Arizona Press.

Milner, A. (1995). *The invention of politics in colonial Malaysia.* Cambridge: Cambridge University Press.

Morsbach, H. & Tyler, W. J. (1986). A Japanese emotion: *Amae.* In R. Harré (Ed.), *The social construction of emotions* (pp. 289-307). Oxford: Basil Blackwell.

Murase, T. (1984). Sunao: A central value in Japanese psychotherapy. In A. Marsella & G. White (Eds), *Cultural conceptions of mental health and therapy* (pp. 317-329). Dordrecht: Reidel.

Ortony, A., Clore, G. L. & Collins, A. (1988). *The cognitive structure of emotions.* Cambridge: Cambridge University Press.

Peeters, B. (1993). *Commencer* et *se mettre à*: Une description axiologico-conceptuelle. [Verbs denoting commencement in French: An axiologico-conceptual description]. *Langue française* [French Language], *98,* 24-47.

Plutchik, R. (1994). *The psychology and biology of emotion.* New York: Harper & Collins.

Renwick, G. W. (1980). *Interact: guidelines for Australians and North Americans.* Chicago: Intercultural Press.

Rogers, M. L. (1993). *Local politics in rural Malaysia.* Kuala Lumpur, Malaysia: S. Abdul Majeed.

Russell, J. A. (1991). Culture and the categorization of emotions. *Psychological Bulletin, 110,* 426-450.

Sapir, E. (1949). *Selected writings of Edward Sapir in language, culture and personality.* Berkeley: University of California Press.

Scherer, K. R. (1994). Towards a concept of "modal emotions". In P. Ekman & R. J. Davidson (Eds), *The nature of emotion: Fundamental questions* (pp. 25-31). New York/Oxford: Oxford University Press.

Scherer, K. R., Wallbott, H. G. & Summerfield, A. B. (Eds). (1986). *Experiencing emotion. A cross-cultural study.* Cambridge: Cambridge University Press.

Shweder, R. A. (1991). *Thinking through cultures: Expeditions in cultural psychology.* Cambridge, MA: Harvard University Press.

Shweder, R. A. (1993). The cultural psychology of the emotions. In M. Lewis & J. M. Haviland (Eds), *Handbook of emotions* (pp. 417-431). New York: The Guilford Press.

Shweder, R. A. (1994). "You're not sick, you're just in love": Emotion as an interpretive system. In P. Ekman & R. J. Davidson (Eds.), *The nature of emotion: Fundamental questions* (pp. 32-44).

SJP (1958-69). *Slownik Jezyka Polskiego* (Dictionary of the Polish Language). Edited by W. Doroszewski, 11 vols., Warsaw: PWN.

Sommers, S. (1984). Adults evaluating their emotions: A cross-cultural perspective. In C. Z. Malatesta & C. E. Izard (Eds.), *Emotion in adult development* (pp. 319-338). Beverely Hills, CA: Sage.

Travis, C. (1998). Omoiyari as a core Japanese value: Japanese-style empathy? In A. Athanasiadou & E. Tabakowska (Eds.), *Speaking of emotions: Conceptualization and expression* (pp. 55-82). Berlin: Mouton de Gruyter.

Vreeland, N., Dana, G., Hurwitz, G., Just, P, Moeller, P. & Shinn, R. (1977). *Area handbook for Malaysia* (3rd ed.). Glen Rock, NJ: Microfilming Corporation of America.

White, G. M., & Kirkpatrick, J. (Eds). (1985). *Person, self and experience.* Berkeley: University of California Press.

Wierzbicka, A. (1992a). *Semantics, culture, and cognition.* Oxford: Oxford University Press.

Wierzbicka, A. (1992b). Talking about emotions: Semantics, culture and cognition. *Cognition and Emotion, 6,* 285–319.

Wierzbicka, A. (1992c). Defining emotion concepts. *Cognitive Science, 16,* 539-581.

Wierzbicka, A. (1993). A conceptual basis for cultural psychology. *Ethos, 21,* 205-231.

Wierzbicka, A. (1994a). Everyday conceptions of emotion (a semantic perspective). In J. Russell, J. M. Fernadez-Doles, A. Manstead, & J. C. Wellenkamp (Eds), *Everyday conceptions of emotion* (pp. 17-47). Dordretch: Kluwer.

Wierzbicka, A. (1994b). "Cultural scripts": A semantic approach to cultural analysis and cross-cultural communication. In L. Bouton & Yamunu Kachru (Eds.), *Pragmatics and language learning* (pp. 1-24). Urbana-Champaign: University of Illinois.

Wierzbicka, A. (1994c). "Cultural Scripts": A new approach to the study of cross-cultural communication. In M. Pütz (Ed.), *Language contact and language conflict* (pp. 67-87). Amsterdam: John Benjamins.

Wierzbicka, A. (1994d). Emotion, language and cultural scripts. In S. Kitayama & H. Markus (Eds), *Emotion and culture: Empirical studies of mutual influence* (pp. 130-198). Washington, DC: American Pyschological Association.

Wierzbicka, A. (1995a). Emotion and facial expression: A semantic perspective. *Culture and Psychology, 1,* 227-258.

Wierzbicka, A. (1995b). Kisses, handshakes, bows: The semantics of nonverbal communication. *Semiotica, 103,* 207-52.

Wierzbicka, A. (1996a). *Semantics, primes and universals.* Oxford: Oxford University Press.

Wierzbicka, A. (1996b). Contrastive sociolinguistics and the theory of "cultural scripts": Chinese vs. English. In M. Hellinger & U. Ammon (Eds), *Contrastive sociolinguistics* (pp. 313-344). The Hague: Mouton.

Wierzbicka, A. (1997a). Japanese cultural scripts: Cultural psychology and "cultural grammar". *Ethos, 24,* 527-555.

Wierzbicka, A. (1997b). *Understanding cultures through their keywords.* Oxford: Oxford University Press.

Wierzbicka, A. (1999). *Emotions across languages and cultures: Diversity and universals.* Oxford: Oxford University Press.

Wierzbicka, A. (2001). A culturally salient Polish emotion: Przykro ['pshickro]. *The international journal of group tensions (Special issue "Emotions in cultural contexts in space and time"), 30,* 3-27.

Wilder, W. D. (1982). *Communication, social structure and development in rural Malaysia.* London: Athlone Press.

Wilkins, D. (1986). Particles/clitics for criticism and complaint in Mparntwe Arrernte (Aranda). *Journal of Pragmatics, 10,* 575-596.

Wilson, P. J. (1967). *A Malay village in Malaysia*. New Haven: Hraf Press.

Wolfson, N. (1983). An empirically based analysis of complimenting in American English. In N. Wolfson & E. Judd (Eds.), *Sociolinguistics and language acquisition* (pp. 82-95). New York: Newbury House Publishers.

– 3 –

Integrating Verbal and Nonverbal Emotion(al) Messages

Sally Planalp and Karen Knie
University of Montana

Everyone knows that verbal and nonverbal cues fit together into integrated messages like interlocking pieces of a puzzle, but nobody really knows how. It is one of the most fundamental and intriguing questions about social interaction, and at the same time it is one of the most intractable. Nowhere is this more apparent than in the study of messages of emotion. We know that emotion is expressed or communicated through words, faces, voices, and bodies, usually in rich combinations, but we know very little about how those cues work in concert or in conflict. Despite decades of prolific and highly sophisticated research on verbal, facial, and vocal expressions in particular (Bowers, Metts, & Duncanson, 1985), we have only a handful of studies that attack the problem of how cues combine, and only a few of them are recent. It is as if researchers have given up studying the puzzle to work on the pieces, and that is understandable considering how complex the pieces themselves are.

Understandable as it may be, many scholars remind us that the pieces are not the puzzle. Jorgensen (1998) said that "in essence, by focusing on one element (i.e., verbal) of the emotional appeal at the exclusion of the other dimensions of the message (i.e., nonverbal), researchers are no longer studying valid communication processes, but rather disassociated parts of the whole" (p. 407). If nonverbal cues modify, augment, illustrate, accentuate, and contradict the words they accompany (often the first thing students learn about nonverbal communication), focusing on the words alone cannot capture their richest and most subtle meanings (Burgoon, 1994). Streeck and Knapp (1992) also observed that, "communication is 'embodied'"…and the various sensory modalities are "inter-organized" (p. 5). And Pittam and Scherer (1993) warned that: "we cannot afford to continue to study single modalities, treating facial, vocal, and postural expression as if they were completely separate domains" (p. 194).

At this point, it may be useful to take a step back to reflect on the different ways in which the pieces might fit together. Our goal is to describe three basic "logics" that explain how and why verbal and nonverbal cues fit together into integrated messages. The three are general models of communication, but can be applied specifically to communicating emotion. We can find evidence for the three models in our everyday and our scholarly terminology, in the assumptions that we make, in the questions that intrigue both scholars and non-scholars alike, and in the ways we try to answer those questions.

MESSAGE DESIGN LOGICS

We begin with a deceptively simple question and a deceptively simple answer. The question is "Why do we communicate?" (or at least "Why do we *believe* that we communicate?"). The answer is: to express ourselves, to understand one another, and to accomplish other goals (primarily social ones). These are the core elements of O'Keefe's (1988) "message design logics" or implicit theories of communication. Although you might think that the three logics were the basis for a deductive analysis of communication, in fact they were derived inductively when O'Keefe and Shepherd observed that people who tried to persuade others by taking multiple goals into account in their messages used three different approaches to their task. Some seemed to say whatever was on their minds, using *language as a medium for expressing thoughts and feelings* (*Expressive Logic*, O'Keefe, 1988, p. 84-87). Others tried to say "the normal and appropriate thing to say in those circumstances," using *communication as a game played cooperatively, according to socially conventional rules and procedures* (*Conventional Logic*: O'Keefe, 1988, pp. 86-87). Still others designed messages to pursue and negotiate mutual goals," *using communication as the creation and negotiation of social selves and situations* (*Rhetorical Logic*: O'Keefe, 1988, pp. 87-88). We see a striking parallel between these three logical types, the ways in which scholars write about emotion messages, and a whole set of assumptions that follow along with them. Sometimes emotion is *expressed* in whatever way it comes out. Why? Just because that's what you are feeling. Sometimes emotion is communicated in ways that are fairly *conventional*. Why? So that someone else will know what you are feeling. Sometimes people communicate emotion more strategically. Why? To accomplish a variety of goals, primarily social goals that involve other people.

Consider, for example, Fernandez-Dols and Ruiz-Belda's study of gold medal winners at the Olympics held in Barcelona in 1992 (Fernandez-Dols & Ruiz-Belda, 1995a, 1995b, 1997). At the time the athletes won the medals, they probably were concentrating on their athletic performances rather than their emotional performances and simply *expressed* their feelings. Interestingly, an observer who was ignorant of the context would interpret the expressions as sadness or anger, not the joy the athletes would be expected to feel. Actually Fernandez-Dols and Ruiz-Belda argued that they were feeling "emotionado" or simply emotional, not the specific emotion of "alegre" (~joy). Later when they

TABLE 3.1

Expressive, Conventional, and Rhetorical Research Paradigms

	Expressive	*Conventional*	*Rhetorical*
Emotion/ communication relationship	Emotion is focus; communication is byproduct	Communication is focus; emotion is content of message	Managing social life is goal; emotion communication is integral part
Metaphor(s)	Container	Playing catch	Dancing
Communicative goal	None	Accuracy	Social coordination
Skill needed	None, unless bottled up	Normal, but some better than others	Potentially highly skilled
Expressions studied	Spontaneous	Posed or spontaneous	Socially situated
Typical model of emotion	Categories	Categories or dimensions	Processes
Verbal/nonverbal link	Link is epiphenomenal; emphasis on nonverbal	Mutually compatible or competing	Multifunctional, flexible
Typical research	Emotion(s) as independent variable; cue(s) as dependent variable(s).	Cue(s) as independent variable(s), accuracy as dependent variable.	Social goal(s) as independent variable(s); emotional message(s) as dependent variable(s)
Research problems	Experimental control, highly individualistic	Issue of what counts as accuracy, manipulating cues	Complexity; no clear boundaries
Research possibilities	Body movement, verbal	Evolving interpretations, understanding, empathy	Effects on social variables

were on the stage receiving their medals, they showed the *conventional* expression of joy that we would expect of gold medal winners—smiling. And, although this was not a part of the research, one might suppose that later behind the scenes they might even have strategically marshaled messages of sadness at others' losses in order to facilitate the *rhetorical* goal of maintaining solidarity with their teammates or Olympians from other countries. All three "message design logics" can be seen in this example and in many other more everyday situations.

Expressive Emotional Communication

From the expressive point of view, emotion is the focus and expressions are simply its by-products or manifestations. The underlying metaphor is emotion pressing out of or leaking from its container, the body. Emotion is *express*ed, re*press*ed, sup*press*ed (press words), or it leaks, bursts, or pours out (like fluids from a container) (Kövecses, 1990). The metaphor comes in three versions—good, neutral, and bad. Expression may be considered good or even necessary because emotions should be "let out" or "vented" rather than "stewing" or "holding them in." Expression might be neutral if emotion cannot be wholly contained and must inevitably leak. Or it might be bad if you have an "outburst" or you "explode." Expressions have no particular purpose, other than to release emotion, and so the only skill involved is keeping it in or letting it out in the right amounts (having a well-working emotional spigot, if you will).

The typical model of emotion that underlies the "expression" model is separate categories of emotion, like different substances in the container. They are hot (anger), warm (love), or cold (fear). Blends are possible, but generally they are less likely than pure substances. In any case, we are better off studying the pure substances because blends are derivative from them anyway (e.g., timid anger). Generally, cues reflect the different kinds of substances (angry cues, joy cues), but they may differ along common dimensions as well (warm glances, cold glances).

Now we come to the central question for this essay: How does expressive message logic lead us to think about the relationship between verbal and nonverbal expressions of emotion? The answer is that they are the different places that emotion leaks or presses out of the container—the body. Emotion presses out through eyes, voice, words, and body movement. Cues come out in combinations only by virtue of being pressed out by the same emotion from the same body (much as steam would come out the spout, the lid, and any other holes in a teapot). Expressing emotion in one modality might facilitate its expression in another, in the same way that gesturing can facilitate verbal encoding of nonemotional messages (Feyereisen & de Lannoy's "coactivation hypothesis," 1991, p. 76; Krauss, Chen, & Chawla, 1996; Rimé & Schiaratura, 1991). For example, crying might facilitate verbal expressions of sadness or pounding one's fist might facilitate yelling (a hypothesis as yet untested, so far as we know). On the other hand, if you block one opening (such as not talking about something), there may be even more pressure forcing emotion out through the face, voice, or body movement (consistent with pressure in the container) (Horowitz et al., 1994). Nonverbal cues might, in fact, be more trustworthy because it's easier to shut off the words than it is the voice, eyes, or body. On the whole, though, we would expect different cues to the same emotion to be consistent with each other; if they are inconsistent, we might suspect that the emotion itself is not pure (perhaps a blend or an odd case).

The typical "expressive" research paradigm is to identify or induce one or more categories of emotion (as independent variables) and to determine what expressions are associated with them (as the dependent variables). In other words, you put different substances in the container and see what is emitted. Ideally, the study would be experimental to optimize both control over what is in the container and the ability to measure what comes out. Context doesn't matter; in fact, the more sterile, the better. And only one person is needed because we assume that people express what they feel, regardless of who else is around.

Research located in the expressive paradigm searches for stable associations between categories of emotion and cues (smiles with joy, furrowed brow with anger, etc), although the dimensions of emotion may also be associated with cues (intensity with vocal volume). Prominent examples are Ekman and Friesen's coding systems for facial cues to emotion (e.g., FACS and its descendants; Ekman & Friesen, 1975), Izard's (1991) work on the emotional expression of infants, and Scherer and colleagues' work on vocal cues (Pittam & Scherer, 1993). Other research topics such as deception detection are also based on the premise that deception provokes emotions such as guilt or "duping delight," which are manifested in "leakage cues" that give away deception (Buller & Burgoon, 1998, pp. 389-391).

Typical problems encountered in this type of research are the following. It may be hard to induce or identify a pure emotion. It may also be difficult to measure expressions, especially several simultaneously (e.g. body movement may be hard to measure with physiological probes in place). More importantly, emotions may overlap or blend (angry fear) and so may expressions (tense smile), challenging the assumption that we can match up each emotion with its distinct expressions. Pittam and Scherer (1993), for example, noted that vocal cues are associated with gradations in arousal, not distinct emotions, although they admit that the jury is still out on whether in time researchers will be able to pin down more exact relationships between categories of emotion and vocal cues.

By stepping outside the assumptions of the expressive approach, one might raise the additional problem that emotional expressions are removed from their natural context. People are not observed expressing emotion while going about their business, especially the business of relating to other people. Instead they are usually alone in a relatively sterile research setting, often responding to recorded materials with no chance that the materials will respond back. If you take the container metaphor seriously, context shouldn't matter, but if you believe that emotional expressions are sensitive to contexts (especially social contexts), unnatural contexts may produce unnatural results.

Within the expressive paradigm, we can point to several possibilities for further research. One is to do more with body movement and actions. Even though it is widely recognized that emotions are expressed bodily, most of the emphasis has been on either the face or on internal physiological cues at the expense of other sorts of body movement (including gestures [e.g., pointing, self-comforting gestures], body orientation [e.g., slumping, strutting, turning away], body agitation [nervous, relaxed], speed of movement [running, lethargy], to name a few). In addition, perhaps because of the power of the "leakage" metaphor, we know more about nonverbal cues than we do about what people are likely to say when they are in specific emotional states, or even how they would describe those states verbally, with the exception of emotion terms (Shaver, Schwartz, Kirson, & O'Connor, 1987) and phrases (Kövecses, 1990). The other chapters in this book are important advances in our knowledge about how emotion is expressed verbally, although most would fit better with either the conventional or rhetorical logics to be discussed later.

From the expressive point of view, verbal and nonverbal cues to emotion merely co-occur with each other and with emotional states, but are not really integrated in any meaningful sense. Thus the expressive models begs the question of how verbal and nonverbal cues go together, although it does provide ways of thinking about multiple cues without having to integrate them. Moving on to the conventional model, we find that integrating verbal and nonverbal cues is a primary concern.

Conventional Emotional Communication

From a conventional point of view, communication is the focus and emotion is the content of the message. The underlying metaphor is one of "playing catch" or "hitting the target." One person sends an emotion and the other person either "gets it" or they "miss each other." Maybe the sender "gets through to" the receiver or perhaps the receiver "is off base." The goal is accurate communication (e.g., Ekman & Oster, 1982), and that requires a fair amount of skill on the part of both the sender and the receiver. And of course some balls are easier to throw and to catch than others are. People's levels of sending and receiving skills vary, but skills can also be developed or trained (as can pitching and catching skills).

The model of emotion that underlies the conventional model tends to be categories (fear, shame, love) or dimensions (usually positive/negative, sometimes intensity). Cues are organized based more or less on body parts (facial, vocal, body, verbal), but also on based on communicative channels that include verbal (transcripts), verbal plus vocal (audiotapes), and all channels (videotapes). The guiding questions tend to center around which cues or channels produce more accurate communication, and if it is inaccurate, which emotions are confused with one another (e.g., anger with fear but not joy) or along what dimension distortion occurs (more negative or less intense than intended). Research on channels is oriented primarily to determining what information is gained by adding channels or lost by taking away channels (comparing videotape to audiotape or audiotape to transcripts).

What is the relationship between verbal and nonverbal messages according to conventional message logic? Let's say that the ball that we are throwing to the other person is made up of a variety of materials. There may be heavy materials (cues) and there may be fluff, so it would be good to tell them apart because you can get along without the fluff. For example, the face might carry the weighty information whereas words are fluff. Perhaps the accuracy with which you can throw the ball depends on the average weight of all the materials, or their total weight. Perhaps some materials interfere with others so the ball is lopsided and hard to throw accurately. In doing research, a number of models for how cues fit together can be compared and contrasted.

The conventional research paradigm is almost exclusively experimental because of the need to control cues and channels as independent variables and the need to measure accuracy as the dependent variable. Typically emotions (anger, fear, joy) are only one factor in a complex factorial design where the independent variables of real interest are types of cues or channels (and interactions among them) and the dependent variable of some form of accuracy. Expressive research moves into a conventional focus when the question becomes whether observers can *identify* certain cues to emotion, with the researchers *presuming* certain associations between emotion and cue because they were pretested or posed. Ekman, Sorenson, and Friesen's (1969) well-known work on cross-cultural identification of facial expressions of emotion would be an example, as would Brownlow, Dixon, Egbert, and Radcliffe's (1997) more recent work on identifying happy dances from points of light located at a dancer's hands, feet, elbows, knees, and so on. Although not strictly experimental, we would include Gottman's (1979) work on the intent-impact discrepancy in marital couples' interactions as conventional research because of its emphasis on accuracy and identifying predictable biases.

The most prototypical genre of conventional research, however, is the multimodal or multichannel study where cues or channels are compared and contrasted to see how they "carry" meaning together. Plausible models are the additive model (e.g., accuracy based on face < accuracy based on face+voice < accuracy based on face+voice+verbal), the averaging model (e.g., positive face + negative voice = neutral message), or the dominant cues model (e.g., visual cues carry more meaning than vocal cues). Much research has been done from this perspective, including the research that generated the oft-quoted but ill-founded formula that communication is "7% verbal, 38% vocal, and 55% facial" (Mehrabian & Ferris, 1967; Mehrabian & Wiener, 1967). (For critiques, including Mehrabian's own disclaimer, see Burgoon, Buller, & Woodall, 1989; Lapakko, 1997.) Such a simple formula would be very handy, but it is completely unjustified, not only by the specific research from which it was derived, but also by the corpus of research done in that vein on the relative communicative value of verbal and nonverbal cues. Krauss, Chen, and Chawla (1996), for example, studied the role of gestures in conversations to conclude that "such evidence as we have indicates that the amount of information conversational gestures convey is very small—probably too small, relative to the information convened by speech, to be of much communicative value" (p. 442). The problem is that either extreme may be justified with certain types of data so that conclusions can vary dramatically.

Even the simplest research on cue and channel combinations (though not simple in any absolute sense) produces incredibly complicated results. Cues do not simply "add up" or "average out," nor does any single type of cue dominate others in any consistent way. In fact, most of the research on cue and channel combinations serves as a tutorial on interaction effects. In one fairly typical study, Wallbott and Scherer (1986) analyzed accuracy using an Analysis of Variance (ANOVA) with four factors: emotion (four emotions), conditions (audiovisual, video, audio, filtered audio), actor (six actors), and situation (scenes that did or did not include a child). Eight interaction effects were significant, five of them at p.<.001 (including the 4-way). In another study, "video primacy" was moderated by affect (+/-), video channel (body or face), sex of decoder, degree of discrepancy, sex of encoder, and judgment dimension (Noller, 1984, p. 140).

The results always seem complex (e.g., Ekman, Friesen, O'Sullivan, & Scherer, 1980; Gallois & Callan, 1986), leading many researchers to conclude that the *meaning* of verbal cues depends on the nonverbal cues that accompany them and vice versa (Krauss, Apple, Morency, Wenzel, & Winton, 1981; Wallbott, 1988). Feyereisen and de Lannoy (1991), for example, concluded their chapter on the "Autonomy of Gestures and Speech" by saying that there isn't any autonomy. Speech influences how signals from the body are interpreted, and what is said is affected by nonverbal cues, often in very complex ways.

The findings themselves are the biggest can of worms for the conventional model of emotional communication, but there are others, both methodological and conceptual. The prevalence of complicated interaction effects leads to the logical conclusion that it is not safe to leave out any possible contributing factor, which in turn leads to bloated ANOVA's. It is simply not possible to include all the factors that are needed to do a truly comprehensive and definitive study or even to compare across studies. Assessing accuracy presents its own statistical problems (e.g., see Russell, 1994), but there are operational ones as well. How does a researcher guarantee that someone is communicating, let's say, anger? You have two choices. You can make him/her feel angry by showing an infuriating movie, taking back the modest reward you promised, or insulting him/her, but still you have no guarantee that you have induced anger. So who is to say that a judge did not astutely pick up some fear in place of or in addition to what the researcher presumed was anger? Your second choice is to ask him/her to pose anger, but we know that posed expressions are different from spontaneous ones (Motley & Camden, 1988).

Taking the conventional model of communication on its own terms, the most serious gap is in verbal cues to emotion, as it was with the expressive model. Perhaps it is not surprising to learn that *the only* verbal cue in the famous 7%-38%-55% formula was the word "maybe." One suspects that degenerate emotional cues may have become a self-fulfilling prophecy. Verbal cues to emotion (in this case liking) are not important, therefore we will ignore them, therefore they become unimportant to research. Verbal messages are something of a nightmare for experimental paradigms, but realism requires that we take them more seriously. In addition, the snapshot "name-that-emotion" approach ignores how even nonverbal cues unfold over time, are often interwoven with complicated verbal content, and result in meaning that is not just accuracy ("That's anger, not fear"), but also understanding and empathy ("I know how you feel" or "I feel how you feel") (Planalp, 1999, chap. 2).

We can also look to existing work within the conventional paradigm for clues about where to go next. Even though Noller's early work (1984) was firmly within the conventional tradition, her findings as well as many others point toward a more complicated model of communication. They showed that when vocal and verbal cues were discrepant in valence (+/-), they took on different duties. "The verbal portion of the message tends to convey attitudes toward the actions of the person being spoken to, while the nonverbal portion of the message tends to convey attitudes toward the person him/herself. The discrepant communication is saying 'I don't like your behavior but I do like you'" (Noller, 1984, p. 136). Walker and Trimboli (1989) also began by exploring channel dominance, but after analyzing specific examples, concluded that "affective messages are sometimes complex combinations of cues in which the roles of the verbal and nonverbal channels are interwoven to communicate several messages simultaneously" (p. 229).

The general failure of mathematical models to yield formulae for combining verbal and nonverbal cues to emotion that are manageable, much less parsimonious, forces researchers to move in the direction of a more complex and situation-sensitive model of communication—the rhetorical model. Combining cues to emotion seems to be more interpretive than it is mathematical, leading many researchers to let people do the combining on the grounds that humans seem to be better at interpreting complex emotional messages than ANOVA models are (Gottman, 1993). The process is probably not magical, but it may just push researchers to a messier and more complicated approach.

Rhetorical Emotional Communication

From a rhetorical point of view, the social situation (including the relationship between the two partners in the interaction) is the focus, and communicating emotion is a way to negotiate that situation. Neither emotion nor communication is foregrounded; rather the two work together (like Escher's writing hands) to accomplish social goals (Frijda & Mesquita, 1991). Social goals are diverse and probably limitless, but the ones cited most often include:

managing one's self-presentation (such as appearing strong, pleasant or even genuine; Clark, Pataki, & Carver, 1996), and managing the relationship between oneself and others (such as communicating warmth; Andersen & Guerrero, 1998a), and managing the other's feelings (such as comforting, aggravating, or goading; e.g., Burleson & Goldsmith, 1998; Witte, 1998). The goals of expression and accuracy may be sacrificed in the interests of other social goals, such as deceiving others about one's true feelings in order to spare them hurt ("I just love your dress!") or using ambiguity in order to avoid taking a stand ("I cannot recommend this person too highly") (Bavelas, Black, Chovil, & Mullett, 1990). Despite the image of rhetoric as one-to-many influence, in dyadic face-to-face encounters, influence is clearly mutual. Finally, whereas the expressive approach emphasized sending emotional messages and the conventional approach emphasized receiving them accurately, both sender and receiver roles are important to the rhetorical transaction.

We propose the metaphor of the dance, with two people coordinating and adjusting to one another often in sensitive and subtle ways. They may follow a standard routine (the tango) or they may improvise their own unique moves. Of course, anybody can bumble through a bad tango, but at its best it is a highly skilled activity, requiring each person to have a complex image of what the other is likely to do, what it means for their own moves, and what sort of dance is appropriate to the setting.

From the rhetorical perspective, it is not clear exactly what form of emotion is communicated. Probably the best overall framework for describing what is communicated is the process model, as described by a number of theorists such as Frijda (1986), Lazarus (1991), Stein, Trabasso, and Liwag (1993), Shaver et al. (1987), and others. According to these models, the emotion process is made up of several components (the exact number depending on the theorist), but usually including precipitating (or eliciting) events, appraisal processes, physiological changes, action readiness (or tendencies), and regulation processes. In its most full-blown and complete form, all components may be communicated, but this is not necessarily and perhaps not even typically the case.

Frijda and Tcherkassof (1997), for example, claimed that what is communicated is not emotion as we would label it verbally (i.e., anger or joy), but rather two crucial components of the emotion process—states of action readiness and eliciting events. If people who are trying to interpret an "emotional" face do not have information about what elicited it, they imagine one. For example, people who were shown slides of facial expressions of emotions (without any other information) responded with comments such as [she looks] "like she sees something very nasty," "she looks the way you look at a small child playing." That is, they imagined situations. Another common type of response was [she is] "shielding herself from someone...or withdrawing from something." That is, they inferred states of action readiness (Frijda, 1953, cited in Frijda & Tcherkassof, 1997, p. 84). If people interpret slides of facial expressions by imagining plausible scenarios that may have provoked them, it is a small leap to assume that if the actual eliciting event is known, it will be taken into account in interpreting someone's response. She looks like she sees something very nasty all right—the casserole over there that is burned to a crisp. The next question is likely to be—what is she inclined to do (action tendencies)? She looks like she's going to yell at me for letting it burn. Oh, oh.

It could also be that what is communicated depends on the goals of both the sender and receiver. If I scream, point, and yell "Landslide!" I am trying to communicate the eliciting event and the urgency of my feelings in order to get you to run in the right direction; my physiological reactions, action tendencies, and coping strategies are largely irrelevant. On the other hand, if I am holding a gun to your head, it is far more important for you to know how aroused I am, what I am inclined to do, and how well I am able to control my emotions than what events precipitated my feelings or how I appraise them.

Not only do the communicative goals of the interactors influence what cues to emotion are sent and received, the process of communicating the emotion may also influence the emotional experience itself. Emotion is not fixed in advance and then either expressed or intentionally communicated; instead, emotions evolve in conjunction with the dynamic social situations in which they are embedded. For example, if you were feeling mildly irritated in a meeting but noticed (even subconsciously) that others backed down when you displayed anger, you might work yourself up into a veritable rage in order to get your way (Bailey, 1983). Similarly, you might show your distress in order to get sympathy from others, but if you find someone who is worse off than yourself, you might try to minimize your own distress in order to help her. Perhaps this is one of the reasons that people who had a job to do during the British air raids in World War II experienced less emotional distress than those who did not (Vernon, 1941).

Ideally, emotional messages would also be studied as dynamic phenomena rather than fleeting expressions that are captured best with a well-timed snapshot. Scherer (1992b) noted that microdynamic analysis of facial expressions shows subtle changes that correspond to some dimensions of the appraisal process (recognizing novelty, pleasantness/unpleasantness, control or lack thereof). It is easy to imagine verbal expressions unfolding in a like manner and being coordinated with changes in facial, vocal, and other nonverbal cues. In addition, situations that elicit emotions may unfold over time. In order to be up-to-date and maximally informative, emotions and their expressions need to change as well. Many emotional episodes such as conflicts are complex and dynamic message exchanges that contain triggers to and expressions of diverse and volatile feelings (e.g., love, hate, guilt, anger, joy, shame, and so on).

What is the relationship between verbal and nonverbal messages according to rhetorical message logic? Because emotion messages are assumed to serve some social function, different kinds of cues provide flexible and subtle ways to influence other people. To use the dance metaphor, an especially long step might set a new pace, eyes gazing in one direction might tell your partner which way to turn, a shift of weight might bring her closer to you. Presumably the cues are consistent most of the time, unless you are unsure what you want to accomplish. But that does not mean that they take their places neatly into categories or dimensions of emotion because some goals might require using complex combinations of cues. Mixed feelings might be communicated using some positive and some negative cues, or sarcasm might be communicated using positive words but a negative tone of voice. In fact, this is very likely because mixed feelings are found very commonly in social interaction (Oatley & Duncan, 1992, p. 264; Omdahl, 1995, p. 158; Planalp, 1998, p. 42), and even if the feelings are not mixed, people may want to equivocate in the interests of "strategic ambiguity" or subtlety (Bavelas et al., 1990). As Frijda and Tcherkassof (1997) noted: "traces of relational activity can be subtle. A tinge of reserve may occur in an expression that is otherwise open and receptive" (p. 101). Or as Buttny (1993) demonstrated: "nonverbal components can be employed to convey messages which would be too threatening to say verbally, such as implied meanings involving challenge, deference, dominance, and the like" (p. 101). Conversely, mixed messages (e.g., happy face with negative verbal content or angry faces saying positive things) are interpreted not as confused (as in the expressive model) or neutral (as in the additive conventional model), but rather as a subtle combination—as insincere, sarcastic, or joking (Friedman, 1979). In any case, "the incongruity is very salient, no matter what the interpretation (p. 465).

Using different types of cues and channels also provides flexibility under changing communicative circumstances. Rimé and Schiaratura (1991, pp. 273-275) reported that listeners may shift their attention from audio to visual cues if audio cues become unintelligible or from verbal to nonverbal cues when nonverbal ones become more intense. It is likely that message producers do the same (Apple & Hecht, 1982), switching their messages from channels that are unavailable to those that are available, although not completely (as can be seen by an angry person gesturing wildly and scowling while talking on the phone). Bavelas and Chovil (1997) postulated "a division of labor among words, gestures, and facial actions so that material is encoded in the most suitable form—for example, personal reactions in faces, shapes and movements in gestures, and abstract categories and syntax in words" (p. 344).

Taking this line of thinking one step further, we might explore how different types of emotion cues interface with components of the process model. It is easy to imagine eye gaze functioning as an indicator of the *object* of emotion (glaring at one person in a meeting) or the *eliciting event* (people looking up at someone standing on a window ledge). Planalp, DeFrancisco, and Rutherford's data (1996) suggest that in naturally occurring situations, vocal cues to emotion often alert observers to something unusual (the *novelty* dimension of appraisal) (e.g., hearing crying upstairs or noticing that your friend is quieter than usual). Vocal cues are good indicators of *intensity* (Pittam & Scherer, 1993), whereas facial cues seem to be better at carrying *valence* information (Russell, 1994). Verbal messages might be needed for complex analyses of *causes* ("I may have offended him, but he's been on edge a lot lately"), other dimensions of appraisal such as coping ("If one more thing happens, I'm going to lose it!"), or action plans ("I think I'll go in and talk to him tomorrow"). Bodily cues might be especially good cues to action readiness (pacing the floor before a big presentation or pulling a loved one closer for a kiss).

Nonverbal expressions of emotion might also be the best medium for accomplishing one kind of goal (hugging to express a desire to be close) whereas verbal expressions might work better for other things (describing a traumatic experience). Keys (1980) noted that "very intense emotions are usually dealt with euphemistically by means of symbolic acts. Direct speech is not the modality for such communications as 'I hate you.' Even 'I love you', said in so many words, can be meaningless. Insulting and obscene gestures replace speech as a means of rejecting and repelling...."(p. 8). Gestures and speech may also stimulate each other, so that acting out a feeling can help make the words come and talking about a feeling can make nonverbal expressions come easier (Rimé & Schiaratura, 1991). The combination of a convincing verbal apology and remorseful nonverbal (facial, bodily) expressions of emotion can also compound the impact of separate cues by communicating that the person *knows* that an apology is called for and *feels* it sincerely (Tavuchis, 1991).

From the rhetorical perspective, we would assume that emotion cues are adapted to their social situations, and it is difficult if not impossible to interpret them out of context. Bavelas and Chovil (1997, p. 335) offered the example of a woman explaining to her conversational partner that even though she likes the food that the other dislikes, she doesn't eat it often. While explaining, she shows a quick disgust face. Taken out of context, one would think that the woman showing the face was feeling disgust herself, but actually she liked the food and was only demonstrating empathy with her partner's disgust. Similarly, Motley's (1993) study of verbal and facial cues to emotion showed facial expressions were interjected into conversations to signal the presence of emotion, but the nature of the affect was implied by context, especially by the verbal context. Motley described the meaning of facial interjections as "something on the order of 'I acknowledge empathy with the emotion suggested by this context'" (p. 25). Walker and Trimboli (1989) also offered two examples of mixed messages serving mixed goals. One is a critic expressing a negative attitude verbally toward the object of criticism (a musical performance) simultaneously with positive affect (humor) directed toward his audience. The other is a TV producer commenting on a weekly budget of $10 by saying with a smile "It's ridiculous." Observers interpreted the message as meaning that he thought the budget was unreasonable but he was proud at having achieved what seemed impossible.

Realistic experimental scenarios may also incorporate the social setting or social goal(s) as independent variables and analyze cues or interpretations of them as dependent variables. For example, Bavelas, Black, Lemery, and Mullett (1986) created a scenario in which someone set a heavy TV down on an apparently injured finger, leading the audience to wince, in part as a private empathic response but also as an expressed sympathetic response. Similarly, Fridlund (1994) studied how the real or imagined presence of other people influenced smiling in response to funny videos, from which he concluded that smiles were both expressive (responses to funny material) and rhetorical (evoked by a real or imagined audience). In both cases, the data indicated that nonverbal cues to emotion were responsive both to felt emotions (empathic distress or amusement) but also to social expectations (expressed empathic distress and shared amusement).

In addition to experimentation, research from the rhetorical paradigm may be based on observation, interviewing, and textual analysis in order to address how social roles influence complex messages. Candace Clark (1990), for example, analyzed "how emotions can serve as 'place claims,' messages about where one wants to stand" (p. 305) in everyday social life. She focused her empirical research (Clark, 1997) on messages of sympathy—how they are justified, and how they both reflect and negotiate gender and power relations between the sympathizer and the sympathizee and implicit social contracts such as norms of obligation and reciprocity. Her analysis was founded on a wide range of research materials, including public statements of sympathy, newspaper articles, and greeting cards, in addition to interviews and observations. Her detailed descriptions make vivid just how impoverished our view of emotional expression would be if we ignored social context.

Moving into more exotic places in the field, anthropologists have studied complex, ongoing verbal and nonverbal exchanges that are used in non-Western cultures to accomplish social goals such as managing conflict. A rich example is Lutz' (1988) study of song-metagu sequences (~justifiable anger-~anxiety/regret) among the inhabitants of Ifaluk, a small Micronesian atoll. The Ifaluk used a varied set of complex verbal and nonverbal cues to let others know when they felt song (~justifiable anger). They gossiped, pouted, refused to speak or to eat with the offending party, refused to eat at all, showed "lantern" eyes, gave declarations of song and reasons for it, threw objects, and in extreme cases threatened suicide or other personal harm (Lutz, 1988, p. 174). The offending party, in turn, shows metagu (~anxiety/regret) by apologizing, some form of payback, or avoidance. No one observing a song-metagu episode could be deluded into believing that these displays *merely* express the *song* and *metagu* that is felt by the participants or *merely* communicate them in a conventional way to insure accuracy (although they may do those things as well). It is obvious that *song* and *metagu* are expressed for social effect as much as for communicative honesty, and they are played to an audience of interested onlookers who may also become actively involved.

By comparison to the expressive and conventional research paradigms, the rhetorical approach seems refreshingly natural and flexible, but commitments to studying dynamic, naturally occurring, and directly observable emotional messages carry their own burdens. For one thing, all of those "conventional" variables don't go away—gender of sender and receiver, power differences, differences between emotions, channel variables, and so on. In fact, in more naturalistic social contexts they all get mixed together into a social soup that must be included as its own complex substance rather than as a simper set of ingredients. More realistic, to be sure; more manageable, definitely not. There is also an additional burden that comes with a commitment to external validity that is not so great within a controlled experimental paradigm where some sacrifice of external validity is assumed in the interests of stronger internal validity. Moreover, if controlled stimuli are hard to construct, uncontrolled stimuli are even harder to select, observe, and record, especially when information about social context is also critical for accurate interpretations.

Research within the expressive and conventional paradigms was also content with a snapshot "affect burst" (Scherer, 1994), whereas the rhetorical paradigm presumes an evolving social situation with both parties in the interaction being mutually responsive to each other (and circumstances) over time. Dynamics are notoriously complex and hard to track, especially when one is trying to capture complex messages where synthesis of verbal and nonverbal cues is assumed rather than studied in its own right. Stopping a videotape and asking observers to explain what they see/hear can interfere with the automatic unconscious synthesis of those cues, not to mention the usual problems with differences between participant and observer perspectives. Moreover, researchers enter with great trepidation into the ghostly realm of intentions and social goals, knowing that they can be myriad, elusive, subtle, and concealed. Perhaps most frightening of all is the recognition that the research has no clear boundaries; there is no safe place from which to stand, point, and say "This is what I am studying, not that other thing over there."

Future possibilities for research within the rhetorical tradition promise even more complexity. In addition to an ever-fluctuating and ever-expanding list of influences on the communicative process (independent variables in experimental research), there is the possibility of a virtually infinite array of possible outcomes or effects (dependent variables in experimental research). Moving beyond the simple accuracy of the conventional approach, we might include in a short list: understanding, empathy, sympathy, various behavioral responses such as reciprocity or compensation (Planalp, 1999, chap. 2), judgments about either or both participants, their relationship, the social circumstances, and an infinite array of judgments suited to a specific purpose such as whether you would vote for, comfort, pay money to, offer a job to, be friends with, believe an excuse of, or follow a recommendation of the target person(s) (for many communicative functions of emotional displays, see Andersen & Guerrero, 1998b). Then we can analyze how a particular emotional display or lack thereof contributes to an evolving social episode. For example, how do Ifaluk people respond when someone fails to express *metagu* as expected, or how do Americans respond to an embarrassing situation (Edelmann, 1987; Metts & Cupach, 1989; Miller, 1996)?

Pursuing research from the rhetorical model is a kind of retreat in the sense that it leads researchers back to the level of detailed, case-by-case description and interpretation of *in situ* verbal and nonverbal message complexes. But in another sense, it is an emancipation from restrictive methods and research contexts and an invitation to study a wide range of public and private messages that have significant emotional and social impact. Moving into uncharted and challenging terrain often requires setting aside sophisticated equipment (if only temporarily) and going back to basics.

PARALLEL UNIVERSES:
SCHERER'S AND GALLOIS' TYPOLOGIES

Other scholars of communication and/or emotion have noticed similar "logics," but with somewhat different emphases and different lines of demarcation between them. Scherer (1992a) has written of vocal expressions of emotion as symptoms, symbols, and appeals, a scheme that closely parallels O'Keefe's, with some exceptions. His term "symptom" closely parallels O'Keefe's "expression" in substance if not in metaphor, although the latter implies more conscious control. Scherer's term "symbol" implies arbitrary cues (although we doubt that he intends this) and puts more emphasis on the cognitive-representational function of emotion cues than does our term "conventional," which emphasizes shared meaning. Scherer's term "appeal" parallels O'Keefe's use of "rhetorical"; in fact, Scherer also referred to the tradition of ancient rhetoric in his discussion. But Scherer emphasized matching via motor mimicry whereas we try to keep the broader emphasis on mutual adjustment. Scherer also believed that "the symbol and appeal functions of nonlinguistic vocal behavior have been much neglected so far" (p. 57), as we believe is true for all other types of cues as well, especially verbal ones.

Gallois (1993) also wrote of three traditions of communication research. The first involves the experience and expression of emotion; the second emphasizes encoding skills, decoding skills, and accuracy; the third explicates rules, codes, and styles used in the communication of emotion. Gallois' distinction is consistent with O'Keefe's and Scherer's, although her review emphasizes the skills and resources required to communicate effectively rather than the reasons for communicating. Her review of individual difference measures of encoding and decoding abilities and of social, cultural, and power-related factors that impact misunderstandings complements our analysis nicely.

Although not described as distinct models of emotional communication, other theorists' conceptions of emotion capture similar processes. For example, Zillman (1996) emphasized precognitive processes that may guide behavioral reactions before conscious monitoring is engaged, although he recognized that if an emotional behavior is disapproved, "excitatory manifestations" or persistent emotional behavior may be "curbed" (p. 247). Note that he used the terms "manifestations" of emotion and "curbed" (consistent with the "expressive" model), but he also wrote that "the modified response may come to dominate the original one," letting people "correct their model of reacting." He left the door open for conventional and rhetorical models, though his own work is based primarily in an expressive one. Similarly, many (if not most) emotion theorists combine something like "spontaneous expressive tendencies" (Buck, 1984, 1989) with something like "display rules" (Ekman & Friesen, 1975) into fused models, if they recognize the distinction at all.

What all of these formulations have in common is that they are founded on continua from the solo to the social, the spontaneous to the strategic, and the simple to the skilled. In addition, they draw on everyday ways of thinking and speaking about communicating emotion as well as on scholarly research traditions.

HOW DO THE LOGICS FIT TOGETHER?

Because of the tripartite terminology, it is easy to assume that communicative reality falls into one and only one box or that researchers must make a commitment to one and only one approach (ignoring or demeaning the others, of course). Instead, we believe that social interaction is so complicated that all three may and often do operate simultaneously. It is possible to express heartfelt distress (expressive) so that others will at least know what you feel (conventional) and perhaps help if they can (rhetorical). It may also be possible to separate and operate from only one logic, but that is more controversial. For example, it may be possible only to express emotion if it is entirely in private (though Fridlund would say expression cannot be separated from communication), or to communicate only so that others would know how you felt (although Frijda & Mesquita [1991] would say rhetorical goals nearly always play a part), or to mislead others for your own purposes (rhetorical only) (we can't think of anyone who would disagree here).

Scherer (1992a, p. 48) believed that all three logics go together even in vervet monkey calls that express fear, denote the presence of a predator, and serve to warn other vervets. Phylogenetically speaking, it is likely that one type of logic builds on another. Expressive cues are probably not even the starting point; rather, they may be founded in the action tendencies that are an essential component of the emotion itself. The disgust face, for example, may be built on functional responses to noxious substances—blocking their smell and extruding them through the mouth. Shame is often expressed as a kind of verbal and nonverbal hiding (evasiveness, cowering posture; Retzinger, 1991), a functional response to escape others' disapproval. Darwin's (1872/1965) analysis of the expression of emotion in man and animals" follows this reasoning. Conventional emotional displays can evolve from an expressive foundation as they become widely recognizable by members of the species. Rhetorical uses, of course, depend on conventional understandings to be effective (crying in order to elicit comfort), in addition perhaps to having some additional logic of their own (e.g., understandings of sarcasm or joking). The greatest challenge to such a neat stairstep model is, of course, patterns of cultural misunderstandings of nonverbal cues, in addition to its obvious inapplicability to verbal cues at the expressive level. As we move further away from the biological and into the cultural, we trade cultural flexibility for inter-cultural misunderstanding.

Scherer (1994, p. 166) also wrote of the "push" and "pull" of emotional communication, the push coming from within the individual (expressive forces) and the pull coming from outside (rhetorical opportunities and constraints). Several empirical findings are consistent with this view. For example, Kraut and Johnston (1979) found that bowlers smiled when they made strikes and spares, but even more so when they looked at their companions. Similarly, Bavelas et al. (1986) found that observers show distress (not amusement) when they see someone appear to smash an already injured finger (the push of empathic emotion), although more often when that person is facing them (the pull of expressing sympathy). Fridlund (1994) found that people showed amusement (not distress) when watching a funny video (the push of felt emotion), but more so in the presence of other people (the pull of being in the same emotional state as others). Certain conditions such as intoxication or overwhelming stress may also give the advantage to "push" forces at the expense of conscious control, just as extreme self-awareness may pull people toward social appropriateness.

Through verbal messages, people may also appeal to one or another logic directly if it suits their purposes. Consider: "I'm just telling you how I feel" (expressing myself even though it may hurt your feelings). Or "I just want you to understand" (accuracy is only goal, though if you chose to help I wouldn't complain). Or "If you really loved me you would know what I want" (you should know my goals without my having to express or communicate them). Displayed so blatantly, all such messages sound strategic and perhaps they are. But in any given case, the primary goal may be more strongly weighted toward "getting it off your chest," or "being on the same wavelength" or "working things out" through mutual adjustment.

Individuals may also differ in the typical message logics they use. Burgoon (1994, p. 236) emphasized that individuals have consistent biases in the cues they rely on in encoding and decoding expressions of emotion. Let's say, for example, that an "expressive" sender focuses on nonverbal cues, a "conventional" sender on verbal cues, and a "rhetorical" sender on verbal/nonverbal combinations. If an "expressive" encoder sends a nonverbal emotional message to a "conventional" decoder relying on verbal cues, the message may not only be lost, but it may lead to misunderstanding, even conflict. Thus, individual cue bias and reliance on a specific message design logic adds even greater complexity and challenge to the situation, not only for the researcher, but for the interaction participants themselves.

CONCLUSIONS

Where does that leave us in addressing the issue of how verbal and nonverbal cues to emotion go together (or fail to) as a part of an integrated message? The three "message design logics" or implicit theories of communication reviewed here give three answers to the question. Cues to emotion may cohere or conflict: (a) because that's the way emotion comes out, (b) because that's the most accurate way of communicating the feeling, and/or (c) because that particular configuration of cues is the best way to accomplish social goals.

So either there are three types of puzzles into which verbal, facial, vocal, body, and other cues to emotion fit together, or the whole puzzle is a kind of three-dimensional Rubic's cube that can only be solved by considering the effects of adjusting one piece on all three dimensions simultaneously. For scholars who value parsimony, this is a frightening thought. Yikes! :-o It is a situation that warrants our collective concern, toward which we (the writers) have strong negative feelings, and about which we are warning you (the readers). Our message is a simple verbal message ("frightening"), a complex one (the problem to which "it" refers as developed in this chapter), a vocal messages (simulated as "Yikes!) and a facial one (simulated as :-o). If effective, our message will draw your attention to the issues, give them a higher priority, and perhaps mobilize our efforts to deal with them. Isn't that what emotion is all about?

REFERENCES

Andersen, P. A., & Guerrero, L. K. (1998a). The bright side of relational communication: Interpersonal warmth as a social emotion. In P. A. Andersen & L. K. Guerrero (Eds.), *Handbook of communication and emotion* (pp. 303-329). San Diego: Academic Press.

Andersen, P. A., & Guerrero, L. K. (Eds.) (1998b). *The handbook of communication and emotion.* San Diego: Academic Press.

Apple, W., & Hecht, K. (1982). Speaking emotionally: The relation between verbal and vocal communication of affect. *Journal of Personality and Social Psychology, 42,* 864-875.

Bailey, F. G. (1983). *The tactical uses of passion.* Ithaca, NY: Cornell University Press.

Bavelas, J. B., Black, A., Chovil, N., & Mullett, J. (1990). *Equivocal communication.* Newbury Park, CA: Sage.

Bavelas, J. B., Black, A., Lemery, C. R., & Mullett, J. (1986). "I *show* how you feel": Motor mimicry as a communicative act. *Journal of Personality and Social Psychology, 50,* 322-329.

Bavelas, J. B., & Chovil, N. (1997). Faces in dialogue. In J. A. Russell & J. M. Fernandez-Dols, (Eds.) *The psychology of facial expression* (pp. 334-346). Paris: Cambridge University Press.

Bowers, J. W., Metts, S. M., & Duncanson, W. T. (1985). Emotion and interpersonal communication. In M. L. Knapp & G. R. Miller (Eds.), *Handbook of interpersonal communication* (pp. 500-550). Beverly Hills, CA: Sage.

Brownlow, S., Dixon, A. R., Egbert, C. A., & Radcliffe, R. D. (1997). Perception of movement and dancer characteristics from point-light displays of dance. *The Psychological Record, 47,* 411-422.

Buck, R. (1984). *The communication of emotion.* New York: Guilford.

Buck, R. (1989). Subjective, expressive, and peripheral bodily components of emotion. In. H. Wagner & A. Manstead (Eds.), *Handbook of psychophysiology* (pp. 199-221). Chichester, UK: Wiley.

Buller, D. B., & Burgoon, J. K. (1998). Emotional expression and the deception process. In P. A. Andersen & L. K. Guerrero (Eds.), *Handbook of communication and emotion* (pp. 381-402). San Diego: Academic Press.

Burgoon, J. K. (1994). Nonverbal signals. In M. R. Knapp & G. R. Miller (Eds), *Handbook of interpersonal communication* (2nd ed., pp. 229-285). Thousand Oaks, CA: Sage.

Burgoon, J. K., Buller, D. B., & Woodall, G. (1989). *Nonverbal communication: The unspoken dialogue.* New York: Harper & Row.

Burleson, B. R., & Goldsmith, D. J. (1998). How the comforting process works: Alleviating emotional distress through conversationally induced reappraisals. In P. A. Andersen & L. K. Guerrero (Eds.), *Handbook of communication and emotion* (pp. 245-280). San Diego: Academic Press.

Buttny, R. (1993). *Social accountability in communication.* London: Sage.

Clark, C. (1990). Emotions and micropolitics in everyday life: Some patterns and paradoxes of "place." In T. D. Kemper (Ed.), *Research agendas in the sociology of emotions* (pp. 305-353). Albany, NY: State University of New York Press.

Clark, C. (1997). *Misery and company.* Chicago: University of Chicago Press.

Clark, M. S., Pataki, S. P., & Carver, V. H. (1996). Some thoughts and findings on self-presentation of emotions in relationships. In G. J. O. Fletcher & J. Fitness (Eds.), *Knowledge structures in close relationships* (pp. 247-274). Mahwah, NJ: Lawrence Erlbaum Associates.

Darwin, C. (1872/1965). *The expression of the emotions in man and animals.* Chicago: University of Chicago Press.

Edelmann, R. J. (1987). *The psychology of embarrassment.* New York: Wiley.

Ekman, P., & Friesen, W. V. (1975). *Unmasking the face: A guide to recognizing emotions from facial cues.* Englewood Cliffs, NJ: Prentice-Hall.

Ekman, P., Friesen, W. V., O'Sullivan, M., & Scherer, K. (1980). Relative importance of face, body, and speech in judgments of personality and affect. *Journal of Personality and Social Psychology, 38*, 270-277.

Ekman, P., & Oster, H. (1982). Review of research, 1970-1980. In P. Ekman (Ed). *Emotion in the human face* (2nd ed., pp. 147-173). New York: Cambridge University Press.

Ekman, P., Sorenson, E. R., & Friesen, W. V. (1969). Pan-cultural elements in facial displays of emotions. *Science, 164*, 86-88.

Fernandez-Dols, J. M. & Ruiz-Belda, M. A. (1995a). Are smiles a sign of happiness? Gold medal winners at the Olympic games. *Journal of Personality and Social Psychology, 69*, 1113-1119.

Fernandez-Dols, J. M. & Ruiz-Belda, M. A. (1995b). Expression of emotion versus expressions of emotions. In J. A. Russell, J.-M. Fernandez-Dols, & A. S. R. Manstead (Eds.), *Everyday conceptions of emotion* (pp. 505-522). Dordrecht, Netherlands: Kluwer.

Fernandez-Dols, J. M. & Ruiz-Belda, M.-A. (1997). Spontaneous facial behavior during intense emotional episodes: Artistic truth and optical truth. In J. A. Russell & J. M. Fernandez-Dols (Eds.), *The psychology of facial expression* (pp. 255-274). Paris: Cambridge University Press.

Feyereisen, P., & de Lannoy, J.-D. (1991). *Gestures and speech: Psychological investigations.* New York: Cambridge University Press.

Friedman, H. S. (1979). The interactive effects of facial expressions of emotion and verbal messages on perceptions of affective meaning. *Journal of Experimental Social Psychology, 15*, 453-469.

Fridlund, A. J. (1994). *Human facial expression.* San Diego: Academic Press.

Frijda, N. H. (1986). *The emotions.* Cambridge: Cambridge University Press.

Frijda, N. H., & Mesquita, B. (1991). *The various effects of emotion communication.* Paper presented at the Sixth ISRE meeting, Saarbrücken, Germany.

Frijda, N. H., & Tcherkassof, A. (1997). Facial expressions as modes of action readiness. In J. A. Russell & J. M Fernández-Dols (Eds.), *The psychology of facial expression* (pp. 78-102). Paris: Cambridge University Press.

Gallois, C. (1993). The language and communication of emotion: Universal, interpersonal, or intergroup? *American Behavioral Scientist, 36*, 309-338.

Gallois, C., & Callan, V. J. (1986). Decoding emotional messages: Influence of ethnicity, sex, message type, and channel. *Journal of Personality and Social Psychology, 51*, 755-762.

Gottman, J. M. (1979). *Marital interaction: Experimental investigations.* New York: Academic Press.

Gottman, J. M. (1993). Studying emotion in social interaction. In M. Lewis & J. M. Haviland (Eds.), *Handbook of emotions* (pp. 475-487). New York: Guilford.

Horowitz, M. J., Milbrath, C., Jordan, D. S., Stinson, C H., Ewert, M., Redington, D. J., Fridhandler, B., Reidbord, S. P., & Hartley, D. (1994). Expressive and defensive behavior during discourse on unresolved topics: A single case study of pathological grief. *Journal of Personality, 62*, 527-563.

Izard, C. E. (1991). *The psychology of emotions.* New York: Plenum Press.

Jorgensen, P. F. (1998). Affect, persuasion, and communication processes. In P. A. Andersen & L. K. Guerrero (Eds.), *Handbook of communication and emotion* (pp. 403-422). San Diego: Academic Press.

Keys, M. R. (1980). Language and nonverbal behavior as organizers of social systems. In M. R. Keys (Ed.), *The relationship of verbal and nonverbal communication* (pp. 3-33). The Hague, Netherlands: Mouton.

Kövecses, Z. (1990). *Emotion concepts.* New York: Springer-Verlag.

Krauss, R. M., Apple, W., Morency, N., Wenzel, C., & Winton, W. (1981). Verbal, vocal, and visible factors in judgments of another's affect. *Journal of Personality and Social Psychology, 40*, 312-320.

Krauss, R. M., Chen, Y., & Chawla, P. (1996). Nonverbal behavior and nonverbal communication: What do conversational hand gestures tell us? In M. Zanna (Ed.), *Advances in experimental social psychology* (Vol. 28, pp. 389-450). New York: Academic Press.

Kraut, R. E., & Johnston, R. E. (1979). Social and emotional messages of smiling: An ethological approach. *Journal of Personality and Social Psychology, 37,* 1539-1553.

Lapakko, D. (1997). Three cheers for language: A closer examination of a widely cited study of nonverbal communication. *Communication Education, 46,* 63-67.

Lazarus, R. S. (1991). *Emotion and adaptation.* New York: Oxford University Press.

Lutz, C. A. (1988). *Unnatural emotions.* Chicago: University of Chicago Press.

Mehrabian, A., & Ferris, S. (1967). Inference of attitudes from nonverbal communication in two channels. *Journal of Consulting Psychology, 31,* 248-252.

Mehrabian, A., & Wiener, M. (1967). Decoding of inconsistent communications. *Journal of Personality and Social Psychology, 6,* 109-114.

Metts, S., & Cupach, W. R. (1989). Situational influence on the use of remedial strategies in embarrassing predicaments. *Communication Monographs, 56,* 151-162.

Miller, R. S. (1996). *Embarrassment: Poise and peril in everyday life.* New York: Guilford.

Motley, M. T. (1993). Facial affect and verbal context in conversation: Facial expression as interjection. *Human Communication Research, 20,* 1, 3-40.

Motley, M. T., & Camden, C. T. (1988). Facial expression of emotion: A comparison of posed expressions versus spontaneous expressions in an interpersonal communication setting. *Western Journal of Speech Communication, 52,* 1-22.

Noller, P. (1984). *Nonverbal communication and marital interaction.* Oxford: Pergamon Press.

Oatley, K., & Duncan, E. (1992). Incidents of emotion in daily life. In K. T. Strongman (Ed.), *International review of studies on emotion* (Vol. 2, pp. 249-293). Chichester, UK: Wiley.

O'Keefe, B. J. (1988). The logic of message design: Individual differences in reasoning about communication. *Communication Monographs, 55,* 80-103.

Omdahl, B. L. (1995). *Cognitive appraisal, emotion, and empathy.* Hillsdale, NJ: Lawrence Erlbaum Associates.

Pittam, J., & Scherer, K. R. (1993). Vocal expression and communication of emotion. In M. Lewis & J. M. Haviland (Eds.), *Handbook of emotions* (pp. 185-197). New York: Guilford.

Planalp, S. (1998). Communicating emotion in everyday life: Cues, channels, and processes. In P. A. Andersen & L. K. Guerrero (Eds.), *Communication and emotion: Theory, research, and applications* (pp. 29-48). San Diego: Academic Press.

Planalp, S. (1999). *Communicating emotion: Social, moral and cultural processes.* New York: Cambridge University Press.

Planalp, S., DeFrancisco, V., & Rutherford, D. (1996). Varieties of cues to emotion in naturally occurring situations. *Cognition and Emotion, 10,* 137-153.

Retzinger, S. M. (1991). *Violent emotions.* Newbury Park, CA: Sage.

Rimé, B., & Schiaratura, L. (1991). Gesture and speech. In R. S. Feldman & B. Rimé (Eds.), *Fundamentals of nonverbal behavior* (pp. 239). Cambridge: Cambridge University Press.

Russell, J. A. (1994). Is there universal recognition of emotion from facial expression? A review of the cross-cultural studies. *Psychological Bulletin, 115,* 102-141.

Scherer, K. R. (1992a). Vocal affect expression as symptom, symbol, and appeal. In H. Papousek, U. Jurgens, & M. Papousek (Eds.), *Nonverbal vocal communication: Comparative and developmental approaches* (pp. 43-60). Cambridge: Cambridge University Press.

Scherer, K. R. (1992b). What does facial expression express? In K. T. Strongman (Ed.), *International review of studies on emotion* (Vol. 2, pp. 139-165). Chichester, UK: Wiley.

Scherer, K. R. (1994). Affect bursts. In S. H. M. van Goozen, N. E. Van de Poll, & J. A. Sergeant (Eds.), *Emotions: Essays on emotion theory* (pp. 161-193). Hillsdale, NJ: Erlbaum Associates.

Shaver, P., Schwartz, J., Kirson, D., & O'Connor, C. (1987). Emotion knowledge: Further explorations of a prototype approach. *Journal of Personality and Social Psychology, 52,* 1061-1086.

Stein, N. L., Trabasso, T., & Liwag, M. (1993). The representation and organization of emotional experience: Unfolding the emotional episode. In M. Lewis & J. M. Haviland (Eds.), *Handbook of emotions* (pp. 279-300). New York: Guilford.

Streeck, J., & Knapp, M. L. (1992). The interaction of visual and verbal features in human communication. In F. Poyatos (Ed.), *Advances in nonverbal communication* (pp. 3-23). Amsterdam, Netherlands: Benjamins.

Tavuchis, N. (1991). *Mea culpa: A sociology of apology and reconciliation.* Stanford, CA: Stanford University Press.

Vernon, P. (1941). Psychological effects of air raids. *Journal of Abnormal and Social Psychology, 36:* 457-476.

Walker, M. B., & Trimboli, A. (1989). Communicating affect: The role of verbal and nonverbal content. *Journal of Language and Social Psychology, 8,* 229-248.

Wallbott, H. G. (1988). Faces in context: The relative importance of facial expression and context information in determining emotion attributions. In K. R. Scherer (Ed.), *Facets of emotion: Recent research* (pp. 139-160). Hillsdale, NJ: Lawrence Erlbaum Associates.

Wallbott, H. G., & Scherer, K. (1986). Cues and channels in emotion recognition. *Journal of Personality and Social Psychology, 51,* 690-699.

Witte, K. (1998). Fear as motivator, fear as inhibitor: Using the extended parallel process model of explain fear appeal successes and failures. In P. A. Andersen & L. K. Guerrero (Eds.), *Handbook of communication and emotion* (pp. 423-450). San Diego: Academic Press.

Zillman, D. (1996). Sequential dependencies in emotional experience and behavior. In R. D. Kavanaugh, B. Zimmerberg, & S. Fein (Eds.), *Emotion: Interdisciplinary perspectives* (pp. 242-272). Mahwah, NJ: Lawrence Erlbaum Associates.

– 4 –

How to Do Emotions With Words: Emotionality in Conversations

Reinhard Fiehler[1]
Institute for German Language

In this chapter, emotions are not regarded primarily as internal-psychological phenomena, but as socially proscribed and formed entities, which are constituted in accordance with social rules of emotionality and which are manifested, interpreted, and processed together communicatively in the interaction for definite purposes by the persons involved. In the elaboration of such an interactive conception of emotionality, the following aspects are treated: the value of emotionality in linguistic theories; emotions as a specific form of experiencing; the rules of emotionality; communication of emotions as transmission of evaluations; practices of manifestation, interpretation and processing of emotions in the communication process; fundamental interrelations between emotions and communication behavior; and methodology of the analysis of emotions and emotionality in specific conversation types. Finally, the developed theoretical apparatus in the analysis of two short conversation sections is elucidated.

EMOTIONALITY IN LINGUISTIC THEORIES

It is a common conception that people's experiences and feelings influence and occasionally even determine their communicative behavior and the course of conversations. Utterances like *His voice was raised in anger* or *It was no longer a reasonable discussion. The two only poisoned each other* express this concept. Linguists, however, have difficulties handling the connections between emotionality and conversational behavior, and more generally, the interrelations between emotions and language. In many linguistic and communication theories, emotionality has no role or no systematic value (e.g, theory of signs, grammar theory, speech act theory, conversational analysis, etc.). Although there have been a series of attempts to develop these theories to include emotionality (e.g., for the theory of signs by the connotation concept, for the speech act theory by

---

[1]Translated by Harold B. Gill, III. Edited by Susan R. Fussell

the explication of the category "expressive speech act" [cf. Marten-Cleef, 1991]), these theories at their core and in their fundamental assumptions do not provide for emotionality.

If the interrelations between emotions and language or communication is taken as the explicit subject of investigation, this very frequently happens on a *theoretical* level, because a fundamental requirement for clarification exists, (e.g., Bamberg, 1997; Battacchi, Suslow, & Renna 1996; Fries, 1996; Konstantinidou, 1997) and/or on a *programmatic* level (e.g., Caffi & Janney, 1994; Dane, 1987; Herrmann, 1987,) without reference to empirical data. The range of questions and methods in these studies is too broadly scattered and heterogeneous to be covered here (cf. the literature surveys of Fiehler, 1990a, paragraph 2.1 and Drescher, 1997, chap. 2). The views regarding emotions in these studies are strongly influenced by theoretical concepts prevalent in the disciplines "responsible" for emotions like anthropology, philosophy, sociology, and especially psychology (for a newer overview of emotion theories in these disciplines cf. Cornelius, 1996).

When an empirical approach to the interrelations between emotions and language or communication is taken, the studies tend to be *experimental* investigations in which test subjects are presented with emotional language content in predefined situations or asked to produce it as a result of a controlled inducing of emotion (Fries, 1991; Tischer, 1993, gives an overview of appropriate investigations in chap. 3). Even when the degree of situation definition and the restriction of the courses of action for the study participants are manipulated to differing intensities according to the directions of the experiment, so that as one pole conversations—induced and performed by instruction—seem to result (e.g., Thimm & Kruse, 1993), nevertheless all these linguistic data are produced for the purpose of their investigation.

Although the literature regarding the connections between emotions and communication/language has achieved a significant scope and an amazing diversification, only a few works have taken emotions in *everyday communication* as their subject and empirically investigated the manifestation, interpretation, and processing of emotions in interaction using natural conversations.

Emotions are a thorny scientific subject for two primary reasons: First, science is dominated by conceptualizations of humans as primarily *purposeful, rational beings*. This conceptualization is prevalent in the most diverse disciplines and theories—beginning in theories of action, moving to interactional theories and right up to cognitive linguistics. Accordingly, scientific theories have likewise tended to regard communication and conversation as cognitively determined, purposeful, rational, and instrumental. Second, the prevailing conceptualization of emotions as *internal–psychological phenomena* makes their linguistic handling difficult.

Increasing demands for both predictability and reliability lead emotionality to appear increasingly socially dysfunctional and lead to an accordingly negative valuation of emotions (Elias, 1981). A picture of the human being emerges from these demands—surely far from the actual state of affairs—in which people are conceived as purposeful, rationally behaving, thoughtful beings—in their communicative and conversational behavior as well. The postulate of isolated linguistic signs in semiotic theory, the separation of denotation and connotation, as well as the reduction of the communication process to information exchange by means of the denotative component of signs, are milestones on the way to theories conceptualizing language and communication without consideration of emotionality. This cognitive and purposeful–rational orientation is recognizable from the central positions of the concepts of "goal" and "purpose" in theories of verbal acting (speech act theory, discourse analysis) and in the limitation of modeling to exclusively cognitive processes. Another example is conversational analysis. Its strict orientation toward the "communicative surface" prevents the explicit consideration of internal psychological processes such as intentions, cognitions, and emotions. As a result, emotions cannot be integrated systematically into theoretical formulations. Even if they are not eliminated completely, emotions can only be included as "a leftover category of the linguistic view of language" (Ehlich, 1986, p. 319).

On the other hand, it is precisely the common understanding that emotions are primarily internal-psychological phenomena that makes their linguistic and conversation-analytic handling so difficult. In contrast, my assumption here is that emotionality has its place in interaction: Only during processes of manifestation, interpretation, and processing of emotions can they be grasped through linguistics and conversational analysis.

Thus one can be interested in emotions from two different perspectives: First, emotions can be examined in the context of the *personal system*. They are then understood as elements of personal interior life. From this perspective, one can ask, for example, what relationship emotions have to other elements of interior life (e.g., cognitions, motivations, dispositions) and how they come to expression. This is the prevailing everyday point of view as well as the scientific perspective.

On the other hand, emotions can be examined as *public phenomena in social situations of interpersonal interaction*. From this perspective, one can examine the function and value of emotional manifestations in interaction, independent of whether the participants also actually feel the manifested emotions. The focus from this perspective is on how emotions are manifested, mutually interpreted, and processed during interaction, and on the *practices* participants use to manifest, interpret, and process emotions. This perspective considers emotions primarily as elements of interaction and emphasizes their functionality for interaction. At the same time, it regards emotions as socially regulated phenomena and stresses their social figuration. "Such a conceptualization focused on the social and particularly the discursive reality of the feelings is, in my opinion, the only one, that opens the problematics for linguistic questions at all" (Drescher, 1997, p. 112).

EMOTIONS AND EXPERIENCE

People experience a multiplicity of internal–psychological processes and states in everyday life, not all of which they would define as "emotions." For example, one can be strained, surprised, curious, fascinated, and so forth. *Experience* and *action* are the two central strands of the personal-environmental reference. Experiencing is a totalizing mode, in which people experience themselves in their relationships with the environment and with themselves. Experiencing results from actions, accompanies actions, and leads to actions.

Emotions and feelings—which are used here interchangeably —are specific forms of experiencing. One can experience annoyance, disgust, and joy, which represent prototypical emotions. But one can also experience irritation, uncertainty, curiosity, tiredness, and hunger, which are not emotions or at least not "pure" emotions. For example, certain cognitive processes play a substantial role in feelings of uncertainty, and physical conditions play a role in feelings of tiredness and hunger. In addition, emotions can be dominant in the experiential process, but they can occur also in various combinations and mixtures with other forms of experience —and this is probably the rule.

RULES OF EMOTIONALITY

The social basis of emotions becomes particularly clear when we consider the rules of emotionality; rules that determine, to a great extent, how people feel, manifest, and process emotions (Hochschild, 1979). Four types of rules of emotionality can be distinguished, each of which regulate the occurrence of emotion at different levels and within different areas: *emotion rules*, *manifestation rules*, *correspondence rules* and *coding rules*.

Emotion rules indicate the type and intensity of feelings viewed as appropriate and socially acceptable within a given type of situation, from the perspectives of both the person concerned and the other participants. As Scherer, Summerfield, and Wallbott (1983) observed, "There seem to be relatively clear cultural expectations as to how appropriate particular emotions and particular intensities of emotion are in particular situations" (pp. 360-361). Coulter (1979) similarly noted that, "Types of situations are paradigmatically linked to the emotion they afford *by convention.* The link is neither deterministic nor biological, but socio-cultural" (p. 133). The *general form* of emotion rules can be stated as follows:

> If a situation is interpreted as type X,
> it is appropriate and is socially expected,
> to have an emotional experience of the type Y.

For example, if a situation is interpreted to involve an irreparable loss, then sadness is appropriate and socially expected. If I am in such a situation in which there has been an irreparable loss, I expect to feel sadness, and on the basis of this emotion rule, my interaction partner expects that I will feel this way and interprets my behavior in this light.

Manifestation rules regulate the type and intensity of emotions that may be expressed in a particular situation, regardless of what emotion one is actually feeling. For example, if a boy's dog is run over he might cry. The expression, *big boys don't cry,* codifies a manifestation rule that specifies that when men are sad, it is appropriate and socially expected that they will *not* display their feelings of sadness by crying.

Correspondence rules regulate the types of emotions and manifestations expected from conversational partners in response to a person's feelings or displays of particular emotions. If, for example, I see that my conversational partner is sad, I should not continue to feel relaxed and merry, at least I should not show it.

Coding rules are conventions that describe and determine which behaviors count as manifestations of an emotion. Thus they pertain to both the behaviors by which a feeling can be manifested and to the indicators in a person's behaviors that enable interactional partners to recognize that he or she is experiencing an emotion.

In essence, in this model of a system of emotionality rules, not only are the expression or manifestation behaviors subject to social standards and conventions, but there are also *rules for the emotions themselves* toward which those participating in the interaction orient themselves. If an individual experience deviates from the rules, an individual or interactive tuning between emotional requirements and individual emotions can take place via *emotion regulation* (Fiehler, 1990a, pp. 87-93). The system of emotionality rules is socially diverse: The rules vary specifically according to roles, gender, situation, and (sub)culture. The rules of emotionality are thus not by any means universal.

COMMUNICATION OF EMOTIONS AS
TRANSMISSION OF EVALUATIONS

Interaction can be understood as a complex hierarchy of tasks that must be fulfilled to achieve certain goals and purposes. Processes of *evaluation* and *statement* always play a role in the solution of these tasks. These evaluations and statements are likewise to be understood as tasks to be solved either individually or interactively. A number of procedures are available to participants for solving the tasks of evaluation. What we commonly call *emotions* or *emotional processes* can be understood as *a specific procedure* for the solution of such tasks of evaluation and statement. Differently formulated: A part of these evaluation tasks is solved on the emotional level.

From a functional perspective, each emotion can be described as an evaluating statement. The following schema can clarify this:

Emotion A is an evaluating statement
about X
on the basis of Y
as Z.

Table 4.1 shows the allocations that are possible for X, Y, and Z:

TABLE 4.1

Emotions as Evaluating Statements

About X	On the basis of Y	As Z
(1) Situation	(1) Expectations	(1) In agreement
(2) Other person	(2) Interests, desires	(2) Not in agreement
Action		
Characteristics		
(3) One's Self	(3) Social norms and morals	
Action		
Characteristics		
(4) Events and circumstances	(4) Self-concept	
(5) Articles	(5) Picture of the other one	
(6) Mental productions		

For example, if I am annoyed at myself because I have knocked over a vase, this can be understood as an evaluating statement about oneself (or an action one has performed) on the basis of one's self-concept (or expectations over my behavior) as not in agreement. If I am pleased with the thought that I will receive a visit tomorrow, then this can be described as an evaluating statement about a mental production on the basis of my desires (or expectations) as in agreement.

If an emotion is communicated in the interaction, this is equivalent to communication of an evaluating statement or, more generally, an evaluation. If the emotion "disgust" is expressed through behaviors (e.g., mimicry, shuddering, vocal characteristics) or by words (e.g., *I am disgusted by this meal*), then a specific negative evaluation—here, of the meal—is communicated to the interlocutor. This same evaluation could, of course, also be conveyed in ways other than by communication of an emotion (e.g., by evaluation descriptors or formulations: *Terrible!* or *I find the meal repulsive,* etc.).

In order to determine the value of emotional communication for interaction, communication must be understood as multidimensional—as more than just the exchange of information by signs. One must assume instead that communication always has at least two aspects of equal standing: the communication of circumstances *and* the communication of evaluations. Evaluations are always communicated as part of any exchange on any topic. Some of this evaluative content is communicated via emotions.

This fundamental evaluative dimension of utterances has been systematically neglected in linguistic conceptions of communication in favor of the information dimension. But the systematic value of emotional communication can only become clear *as special form of communication of evaluations* when both the evaluative and informative dimensions of communication are taken into account.

COMMUNICATION OF EMOTIONS IN THE INTERACTION

With respect to emotions in interaction, participants solve specific *communication tasks* by means of specific *communicative practices*. Three broad classes of communication tasks can be distinguished: the *manifestation of emotions*, the *interpretation of emotions,* and the interactive *processing of emotions*.

During the course of an interaction, emotions can be *manifested* and communicated to partners through various patterns of behaviors and utterances (especially phenomena of emotional expression and verbal thematizations of emotions), independent of whether the emotions are actually present or not.

In particular, if experience was displayed in an interaction-relevant manner, but also independently of this, the emotional presence is *interpreted* in interaction situations more or less intensively mutually. This emotion task does not necessarily require communication tasks. An interpretation may be made privately, but the results of this interpretation can become relevant to the interaction. An interpretation may also involve communicative sequences in the form of questions (e.g., *Are you angry?*), projective experience thematization (e.g., *Don't be so sad*), or negotiations.

Once an experience or an emotion has been established as an interactive "fact" through manifestation and interpretation, it can be *processed* communicatively. Strategies for processing emotions include "entering," "analyzing," "calling into question," and "ignoring."

Manifestation of Experiencing and Emotions

People can use two different strategies to communicate emotions and experiences: They can give them expression in different ways or they can make them the explicit topic or subject of the interaction. Thus, we can distinguish between practices for the *expression* of experiences and emotions and practices for the *thematization* of experiences and emotions.

In thematization, an experience or an emotion is made the topic of the interaction by a verbalization. Expression, however, is not limited to verbalization (although expressions may naturally accompany verbalizations). Emotional expressions also do not necessarily make experiences although they are manifested as the topic of the interaction. A splenetic *Can you perhaps be on time sometime?* certainly expresses an emotion by means of speech rate, intonation contour, vocal characteristics, and so forth. However, it does not thematize the experience, as would be the case with *Your perpetual lateness really annoys me!* In the latter example, emotional experience is explicitly made the topic of verbal communication, allowing people to communicate about emotions. This is the essential structure of *thematization* of emotions and experiences.

Most often, however, the topic of verbal communication will be something other than emotion, but*besides and at the same time* people communicate emotions by the manner in which they communicate about the topic. The emotions function as evaluating statements with respect to the topic, as well as with respect to further aspects: to other persons, their actions, ourselves, and so on. This is the essential structure of the *expression* of emotions and experience.

Expression

The communication of emotions consists to a high degree of *expressions* of emotions and the interpretation of these expressions. Emotional expression, as it happens at a certain place in the interaction, is a function of *underlying emotions* on the one hand and, on the other hand, of *display rules* specifying what expressions are socially appropriate and expected in a given situation. Emotional expression is thus understood as not exclusively a consequence of internal emotions (this is the usual view), but as determined *equally* from internal emotions *and* manifestation rules.

By *emotional expression* I mean all behaviors (and involuntary physiological reactions) in the context of an interaction that are manifested by a participant with the awareness that they are related to emotions and/or that are perceived and interpreted by the interaction partner accordingly. In this way, emotion expression is conceptualized from the outset in terms of its communicative function within interaction.

The relationship between expression and emotion can be determined on the following basis: In specific situations, specific behaviors and physiological reactions of individuals or groups are understood by other individuals or groups as the expression of a certain emotion of a certain intensity. These expressions may or may not actually express such an emotion. For this determination, the following considerations are relevant:

- Emotion expression, or that which is considered as such, can vary by situation and by person or group.
- Interacting persons do not always interpret emotion expression as a sign of "real" emotions.
- Emotion expression has a conventional aspect.
- Emotion expression is a complex phenomenon, usually consisting of more than one behavior or physiological reaction.
- Any behaviors and physiological reactions can be interpreted by a recipient as emotion expression—independent of how the acting person understands these behaviors and reactions.

Thematization

We can distinguish at least four practices in the *thematization* of experiences and emotions: (a) verbal labeling of experiences and emotions, (b) description of experiences and emotions, (c) designation or description of the events and circumstances relevant to the experience and (d) description or narration of the situational circumstances of an experience.

Verbal Labeling of Experiences and Emotions. Emotions can be thematized by verbal experience labels. Terms for emotions are socially preformed interpretative possibilities for personal experiences; they are socially standardized possibilities of typing and of defining an experience. The whole of these designations forms the *emotion vocabulary or lexicon* of a language. Emotion terms exist at both the general (*feeling, mood, experiencing*) and specific (*fear, joy, fascination*) levels. They are present in nominal, verbal, and adjectival form. The emotion vocabulary cannot be clearly delimited, due to the fact that most expressions we use for internal states also have cognitive, evaluating, motivational, and physiological components of meaning, with which they can become relevant in communication (Fiehler, 1990c).

Descriptions of Experiences and Emotions. Descriptions of experiences and emotions are more or less detailed attempts to clarify a specific experience to the interaction partner by rewritings. Important linguistic means for the implementation of descriptions of experience include, among other things, the use of (a) experiential declarative formulas, (b) frozen metaphorical idioms (phraseologies), and (c) the metaphorical use of expressions:

Experiential declarative formulas are expressions that define what occurs in their scope as experience or emotion. Examples of these expressions include: *I felt (myself) X; I had the feeling X; to me, it was X,* and so on. The scope of these formulas can include experience-designating terms (*I felt anxious/depressed/happy/etc.*), short comparisons (*I felt empty/put upon/etc.*) and comparisons or images using *like* or *as if* (*I felt like the sun king/as if the earth had slid away under my feet/etc.*). *Frozen metaphorical idioms* are conventional figurative-metaphorical expressions for emotions or experiences, such as: *Es kocht in mir (it boils within me); Das haut mich aus den Schuhen (that knocks my socks off); Du treibst mich auf die Palme (you are driving me nuts),* and so on. *Metaphorical use of expressions* refers to other, nonconventional figurative uses of expressions to describe an emotion or experiences, such as:: *Ich hänge durch* (I am sagging); *Die Prüfung steht mir bevor* (my exam is impending); *Ich war völlig zu* (I was completely blocked) and so on.

If one analyzes the full range of figurative language used to describe experience (experience-declarative formulas with short and developed comparisons, fixed metaphorical idioms, metaphorical use of expressions, etc.), one can derive fundamental everyday conceptualizations of emotions which structure and determine our understanding of emotions and their functioning (Kövecses, 1990). Except by labels, experience and emotions can hardly be talked about except by representing them in analogy to other (more concrete) areas (Lakoff &Johnson, 1980). Frequently these are not single analogies, but rather, certain domains are the source for a multiplicity of figurative descriptions of experience. I briefly mention some substantial and productive domains.

Negative experience is frequently conceptualized as PHYSICAL VIOLATION or disturbance of the physical *integrity* (something *hurt me terribly, left a wound, scars, gnaws at me, I shred myself, something hits me like a shot to the heart,* or *goes under the skin*). Furthermore emotions are conceptualized as SENSORY PERCEPTION (*that left a bad taste in my mouth, stinks, I can't stand the smell, it doesn't scratch my itch*). Further concepts are HEAT and COLDNESS (*that left me cold, it boils within me*), the *pressure* rising to the explosion (*it tore my heart, I exploded*) and *taking off from the ground* (*who will fly up in the air*). Close relationships exist between these concepts, and they spill over into concepts used to describe emotional dynamics: RISING AND FALLING OF WATER (*a feeling of fear flooded over/through me, then it ebbed*), INFLAMING AND BURNING, as well as WIND AND STORM.

Positive feelings are conceptualized as, among other things, DECREASE IN WEIGHT, which makes it possible to float or fly (*it took a load off my mind, what a relief, I floated as on clouds, felt like I was in seventh heaven*). This is a special case of conceptualizations that use the dimension *height/depth*, whereby positively evaluated feelings are high, and negatively evaluated are deep (*I am down, torn down to the ground*). The investigation of which conceptualizations are the basis of the descriptions of emotions, how they are connected with more general conceptualization practices and which mixtures or overlays of metaphoric domains are possible, results in a most interesting way of explaining the everyday understanding of emotions.

Labeling/Describing Events and Circumstances Relevant to an Experience. By labeling or describing events or circumstances that have clearly negative or positive consequences for the speaker, experience connected with these events can be made the topic of the interaction. This is particularly true if the utterances are accompanied by appropriate expression phenomena, for example, *My dog was run over yesterday!* (with appropriate mimicry and intonation).

Description/Narration of the Situational Circumstances of an Experience. One's own or other's past experiences can be thematized, as the situational circumstances or the flow of the events are described, reported, or told. They are reported or told, *in order to* clarify an experience in the situation concerned. The aim of such reports or narrations in experience-thematizing intention is to have the listener call the narrated situation to mind and, resorting to the emotion rules valid for this type of situation, grasp how the other one felt in that situation. The description of the situational circumstances and order of events can thereby consist exclusively of the playback of actions and cognitions of the persons involved in the situation concerned. In more complex experience thematizations, verbal designations and descriptions of the experience may be used in addition to descriptions of the situational circumstances. I speak of *a complex experience thematization*, if the speaker uses more than one of the practices I have specified for thematizing a particular experience.

Elements in Thematizations of Experiences and Emotions

The thematization of an emotion can focus on different aspects of the experience, including:

1. the carrier of experience: *P*
2. the type of experience: *E* (for unspecified experience), *A* (for a specific emotion)

3. the intensity of experience: I
4. the dynamics or process of experience: D
5. the object or point of reference of experience: O
6. the bases of experience and the yardsticks of the evaluating statement: G

One can easily recognize in these aspects the definition given above that experience / emotion E is an evaluating statement (specific according to type A, intensity I and dynamic/process D) of a person P to something O with the statement grounded on a specific basis G. However, thematization can also be determined by a number of further aspects:

7. the initiator or the reason of the experience: V
8. the localization of experience in the body: K
9. the appearance of expression and somatic-physiological effects of the experience: AUS
10. the consequences of the experience: F

With the help of these aspects of focusing, we can completely analyze all utterances that thematize an experience on the basis of the practices of verbal labeling or description. First the analysis of two examples of *verbally labeling experience thematizations*:

A. "I   was in complete despair."

P   I   A

B. "He had a disgusting feeling in his gut because of the discussion."

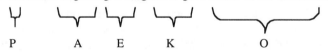

P   A   E   K   O

Now three examples of the analysis of *experience descriptions*:

C. "You are gradually getting on my nerves."

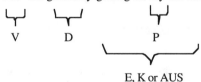

V   D   P

E, K or AUS

D. "It boils inside me."

E I AUS   K   P

E. "That was a considerable shot to the kidneys."

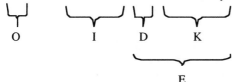

A single experience-thematizing utterance does not need to focus on every aspect mentioned, but it will contain only these aspects. Many different combinations of aspects can be focused, and with different weights and different levels of precision.

Experience thematization can be produced with a large variety of figurative expressions as was demonstrated earlier and as the examples again show. These figurative thematizations of experience can also be analyzed with the help of the ten aspects of focusing specified previously. Figurative expressions (in the context of experience thematization) are always interpreted as the focusing of a specific aspect or a combination of these aspects. The fact that the figurativeness always refers to this relatively small number of aspects protects its comprehensibility, on the one hand, and makes its variety possible, on the other hand.

Manifestation Areas

Interacting individuals manifest their experiences and emotions—whether by expression and/or by thematization—within every area of their behavior. The manifestations can be assigned the following areas, whereby manifestations in the verbal area are somewhat more exactly differentiated:

1. Physiological manifestations (e.g.. trembling, paling).
2. Nonvocal nonverbal manifestations (e.g.. mimic, gesturing, body attitude).
3. Vocal nonverbal manifestations (e.g., affect sounds, laughter, groaning).
4. Paralinguistics (e.g.. voice characteristics, speech rate).
5. Verbal proportion of utterances
   5.1. Manifestation in the linguistic-content form of the verbalization (e.g. word selection).
   5.2. Content-thematic adjustment of the verbalization
      5.2.1. Emotional-verbal utterances e.g. proclaims.
      5.2.2. Verbal-emotional utterances e.g. reproaches, disciplining.
      5.2.3. Verbal labeling / description of experience-relevant events/circumstances
      5.2.4. Description/narration of the situational circumstances of an experience

5.3. Verbal thematization of experience
   5.3.1. Verbal labeling of experience
   5.3.2. Description of experience
6. Conversational behavior
   6.1. Topic (e.g., selection of a sad topic).
   6.2. Type of conversation (e.g., silly jokes).
   6.3. Conversational strategies (e.g., demonstrative denial, shameless openness).
   6.4. Organization of conversation (e.g., overlap, interruption).
   6.5. Conversation modality (e.g., engaged, loosely, ironically).

This system shows how broad the spectrum of the phenomena is by which emotions can be manifested. They distribute themselves over the entire spectrum of communicative behavior.

Interpretation of Experiences and Emotions

Interacting individuals interpret others' experiences and emotions all the time, although with changing intensity and accuracy. The interpretation consists of the fact that (a) specific experiencing is imputed to the other person—also independent of indicators present, that (b) behaviors and physiological reactions are interpreted as emotion expression, and that (c) experience thematization is interpreted. The ascription of an emotion is *a result* of these three components.

In the majority of situations, experience is imputed more or less differentially to the interaction partner. Even when no emotional indicators are present or perceived, emotions can be imputed to others on the basis of (a) the emotion rules that apply to the specific situation, (2) the projection of one's own emotional disposition on the other, and (3) potentially, knowledge about one's partner's emotional disposition.

During an interaction, a partner's behaviors and physiological reactions may be interpreted as an expression of his or her experience and emotions. When other's reactions are interpreted in such a way, I call them *experience and emotion indicators.* For the interpreting person, they are indicators of certain emotions within the situation concerned, but they are not indicators in a general and objective way. In principle, *all* of one's partner's behaviors and physiological reactions can serve as possible forms of an expression of experience and emotions and thus used to interpret their emotional state.

Finally, the interpretation of experiencing also supports itself by the interpretation of the experience thematization of the other person.

Processing of Experiences and Emotions

Once experiences and emotions are established as common fact for the participants by manifestation and interpretation, they can be handled or processed in the interaction. We can distinguish analytically among four processing strategies: (a) "entering" refers to all strategies with which the interaction partner accepts the displayed emotion as appropriate and handles it with expressions of sympathy; (b) "analyzing" refers to strategies by which the suitability of the manifested emotion in terms of intensity and/or type is problematized; (c) "calling into question" refers to strategies by which displayed emotions are not accepted as appropriate; and finally (d) "ignoring" refers to strategies by which the interaction partner—despite having perceived and interpreted the emotion—consciously and obviously avoids acknowledging it and dealing with it interactively in manifest way. The demonstrative character of the avoidance process differentiates between "ignoring" and "passing over." These four strategies can also occur in combination. It is obvious that the three first strategies mentioned are tied to verbal communication processes.

A range of possibilities of the interactive handling of emotions has developed into communicative patterns of action. Patterns of action are socially standardized and conventionalized practices that serve to realize specific functions and purposes that frequently return during social processes (Ehlich & Rehbein, 1986). So, for example, a "sympathy-pattern" is central to the "entering" strategy (Fiehler, 1990a, pp. 150-156), whereas a "divergence" pattern of action—with which a manifested emotion becomes an object of negotiation—is central to "analyzing" and "calling into question." I illustrate the divergence-pattern by a (fictitious) example:

A: C could have saved himself his silly remarks!
B: Don't be so annoyed at yourself. It isn't worth it.
A: Let me, anyway. I want to excite myself at times.

In this example, an emotion expressed by A is called into question by B. B interprets A's expressive phenomena (e.g. word selection, intonation) as annoyance. B's suggestion in the context of the negotiation that starts with his utterance is directed toward a modification of the intensity of experience ("so"). The suggestion for the modification of the experience is implemented in the form of a direct request. The reason recurs due to an emotion rule, according to which a high intensity of experience is coupled to a high importance of the event, which does not exist in B's opinion. The correcting function of the suggestion is predominant. B's attempt to interactively regulate emotion is rejected by A, however, who then uses another experience label ("excite").

This example illustrates that emotions can very probably be "negotiated" and that this negotiation can concern not only the manifestations but also the emotions themselves. Such negotiations are delivered on the basis of implicit or explicit mentioned emotion rules. Feelings are thereby argued on the basis of conceptions about emotions such as whether they are appropriate or inappropriate, justified or unjustified, and so forth. Negotiations of this type make it clear that feelings are by far not a private thing but are socially standardized.

CONNECTIONS BETWEEN EMOTIONS AND COMMUNICATION BEHAVIOR

Thus far, we have considered the tasks that take place during the handling of emotions in interaction from a *process* perspective—that is, from the point of view of the interacting partners—with particular focus on how partners solve emotions tasks in and through communicative processes. Now, we will take an "interactional analysis" perspective—the perspective of a person analyzing a previously documented interaction—and consider what phenomena in the conversation can be used to detect that emotions were manifested, interpreted, and processed by the participants. Phenomena which appear from a process perspective to be, for example, *manifestations* of an emotional experience, will appear from an interactional analysis perspective, if it considers the results of an action, as the *effects* of the emotional experience on the utterances, or, more generally, on communicative acting. The emotional experience affects and modifies the communicative action and is thus evidenced there.

Connected with this idea is the question regarding *fundamental systematic connections* between emotions and communication behavior. Two such general connections can be identified: First, *emotions modify a communicative behavior*. Here, we can distinguish between cases in which an emotion accompanies a communication behavior (which would have taken place without the presence of the emotion) and *affects* the form of the message, and cases in which the emotion *motivates* a communicative behavior and thus is the cause for the utterances, which would not have taken place in such a way without it. Second, *communicative behavior modifies emotions*. In this second connection, we can further distinguish between cases in which the communication behavior influences *one's own* emotionality and cases in which it influences the emotionality of the *interlocutor(s)*. In the first case, verbal techniques of personal emotion regularization are involved.

Three basic assumptions are prerequisites for these considerations of different connections between emotional and communicative behavior. They are explicitly identified here because they are not unproblematic and are not necessarily shared (even though they are constitutive for the common understanding of the connections between emotions and communication; cf. Fiehler. 1990a, pp. 163-167 for the problems of these assumptions):

- Emotional experience and communicative behavior can be understood as entities that are, in principle, *independent* from each other.
- The first assumption implies that emotions can be the *cause* or the *reason* for communicative activities, and also, in contrast, that communicative behavior can be a cause and reason for emotions.
- The third assumption means that there can be an unemotional or *emotionally neutral* mode of communicative activities and that this is the basic mode. If emotions occur, then they modify this neutral way of communication in a recognizable and specific way. According to this view, communicative processes are built up from unemotional (neutral, objective, calm) passages and emotional phases, in which the communicative behavior is affected and modified more or less strongly by emotions.

If one asks where in communication behavior emotions work themselves out or to which phenomena must one pay attention, one can orient oneself to the manifestation areas specified above. In each of these areas, phenomena can be identified that are brought into connection conventionally with the working of emotions and by the presence of which it can be considered that with their causation, emotions were in play or that they are indicators for emotions (cf. Fiehler 1990a, pp. 168-174).

METHODOLOGY OF EMOTION ANALYSIS

Insofar as it is interested in the interdependencies between communication and emotions, the analysis of conversation involves examining these connections in conversation recordings and transcripts. The general questions of such an emotion analysis are: (a) Which processes of manifestation, interpretation, and processing of emotional experience can be found in a conversation? and (b) How are these processes presented in the participants' communication behavior? Because it is not evident during recordings of natural conversations whether and which emotions play a role there, and emotionality usually only is thematized to the smallest extent, an explicit methodology of emotion analysis is of central importance for such investigations. In the following, a six-step methodology is sketched out that makes it possible to ask the questions mentioned, at least to some extent.

The *first* step of the emotion analysis consists of determining what is *thematized* in emotions and emotion interpretations in the participants' interaction. Indicators for thematization include explicit experience labels and descriptions. On the basis of this thematization, one can examine whether specific emotion indicators corresponding to these thematizations can be recognized.

The target of the *second* step is to realize, on the basis of empathy, which emotions and which experiences can play a role for the participants in an interaction of the type concerned and in the specific situation. This step makes the *scope of expected experience* of the participants explicit on the basis of emotion rules.

In the *third* step, the interaction is examined for *indicators* by which emotions and emotion interpretations express themselves. So that this step does not go over the line, it is limited first to sequences of the interaction that are particularly "emotional."

In the *fourth* step, the selected sequences are examined *utterance by utterance* in view of both of the types of systematic connections between emotions and communicative behavior previously specified. That is, it is now sequentially examined as to whether and how the imputed emotions modify the participants' communicative behavior and whether and how this communicative behavior modifies the participants' emotional states.

In the *fifth* step, the analysis, which in the third and fourth steps focused only on selected sequences, is expanded to the entire interaction.

The *sixth* and last step of the analysis pursues another direction. It examines whether communicative patterns of emotion processing can be found in the interaction (e.g., the sympathy pattern, the divergence pattern). This analysis step can also be executed independently of the others.

EMOTIONALITY IN SPECIFIC CONVERSATION TYPES

In the following, I concentrate on investigations that examine emotionality in natural conversations on a interpretive and conversation analytic basis.

It is obvious that attention was directed first toward such conversation types, which according to the everyday understanding are "emotionally pregnant," particularly disputes or therapeutic communication. On the basis of authentic conversations and transcripts, very different investigation questions are possible. On the one hand, in the sense of a case analysis, the reconstruction of the emotionality of a single conversation can be the aim, on the other hand, systematically limited questions can be processed:

- Incidents and processing of a certain experience or a certain emotion (e.g., embarrassment [after errors in instuctions; Brünner, 1987], feelings of indebtedness [Vangelisti, Daly, & Rudnick, 1991], astonishment [Selting, 1995], rage [Christmann & Günthner, 1996], indignation [Schwitalla, 1996], edginess [Hartung, 1996])

- Incidents of specific means or forms of the manifestation (e.g., exaltation [Kallmeyer, 1979a], emphasis [Selting, 1994], interjections and reduplications [Drescher, 1997]).
- Incidents of specific communicative patterns and forms of emotion processing (e.g., sympathy pattern [Fiehler, 1990a; Schwitalla, 1991], tone of appeasement [Schwitalla, 1997]).
- Specifics of the manifestation, interpretation and processing of emotionality in a certain type of conversation, and so on.

Among the particularly emotionally pregnant conversation types are *conversations between physician and patient* (cf. Bliesener, 1982; Bliesener & Köhle, 1986; Fiehler, 1990b; Gaus & Köhle, 1982; Lalouschek, 1993; Löning, 1993; Lörcher, 1983). This type of conversation is characterized by a special problematic concerning the manifestation of experiences and emotions by the patient. Although diseases are frequently connected with intense experience, the somatically oriented medical field expects the patient to keep his or her emotions out of the situation and to limit his or her participation to cooperatively and materially supporting the physician's anamnestic, diagnostic, and therapeutic measures. This frequently succeeds only partially with the patients, however, so that a processing of the manifested emotions then becomes necessary within the physician-patient conversation. This emotional processing often ends unsatisfactorily.

Physicians or medical personnel frequently carry out "feeling work" (Strauss, Fagerhaugh, Suezek, & Wiener, 1980) during the treatment as a purely instrumental function to ease their own work. "Feeling work" refers to dealing with a thematized or anticipated experience of the patient "in the service of the main work process" (Strauss et al., 1980, p. 629).

Therapy conversations are also emotionally intensive in a special way (e.g., Coulter, 1979; Fiehler, 1990a, pp. 239-247; Käsermann, 1995). One of the central goals in such interactions as client-centered therapy is the *verbalization of the contents of emotional experience*. Accordingly, processes of thematization, interpretation, and processing of experience and emotions play a substantial role in therapeutic conversations of this type: The purpose of the therapy sessions is to bring the client's emotions into focus. On the one hand, the client's emotions connected with the *current therapy situation* are processed. On the other hand, the client's *past experiences* are brought into the therapy situation by narrations and reports. The client's *manifestations* and *thematization* of experiences in client-centered therapy differ quantitatively but not qualitatively from those in other conversation types. They are much more frequent and are more developed and differentiated (e.g., complex experience thematization, experience declarative formulas with developed comparisons), and thematizations are over-represented in comparison with the expression of emotions (cf. the following analysis). Qualitative differences, however, exist in the *interpretation* of experience manifestations (by the therapists) and in the *processing* of manifested experiences (by client and therapist).

One of the therapist's central actions is to redirect attention away from the client's reports, narrations, and reflections, to the experience dimension of the reported events. For this, the therapist's most important tools are the focusing and interpretation of experience. During *experience focusing* the therapist does not formulate interpretation, but rather motivates the patient to explore and verbalize his or her experience . In addition, the therapist can formulate an *experience interpretation* more specifically by using a label or description with a hypothetical or determining character. With such experience interpretations three things can occur: they can be ignored, the client can make them his or her own, or they can become an article of a negotiation. The *experience negotiation* can thereby refer to both the intensity and type of experience imputed by the therapist. These differences regarding manifestation, interpretation, and processing of experience and emotions are a direct reflection of the therapeutic theory, which is made operational in the conversation with the client.

A third type of emotionally pregnant communication is *disputes* (Fiehler 1986, 1990a; Kallmeyer, 1979b). If different views, opinions, and interests meet, if a diversity of opinion, a controversy, or a conflict is present, then, for the participants, this is frequently connected with emotions, which manifest themselves in the interaction. However, usually the experience of the participants is not the topic of the interaction but rather is manifested by expression parallel to the argument about another topic. A prerequisite for conflicts is that a contrast exists so that a *position* and an *opposite standpoint* are interactively built up and negotiated. This can cover a pair of utterances or long interaction sequences. The existence of an opposition means that the positions are mutually evaluated as not compatible or contradictory and that on the basis of these evaluations, the opposite standpoint is formulated. In this manner, each contribution for delivering of oppositions contains *components of evaluation*. This is essential for understanding the role of emotions in the delivery of oppositions: A part of these evaluations is communicated by the expression of emotions .

A final example of emotionally intensive communicative activities are *narrations* (e.g., Bloch, 1996; Fiehler, 1990a; Günthner, 1997). The cause for many narrations is that a person has felt particularly strong emotions or has had unusual experiences that were associated with intense emotions. The purpose of the narrations is to clarify these experiences or emotions for another person. This clarification typically takes place in narrations less by means of experience labels or descriptions, but more by description of the situational circumstances and the flow of the events whereby the emotions concerned can and must be inferred on the basis of emotion rules.

This short overview shows that the empirical analysis of emotionality in natural conversations is still in its beginning.

EXAMPLE ANALYSIS

To illustrate the statements regarding therapy conversations in the previous paragraph and to show that empirical data can be analyzed by means of the conceptual instruments presented here, I now examine two short exemplary transcriptions from therapeutic sessions. The following transcription originates from the initial phase of such a session. At the beginning of a session, it is typical for clients to report events relevant to experience from the period between the sessions. This also occurs here. The transcription is segmented with indicated lines, so that the analysis can be followed more easily.

(1) Giving and taking 10: Transcript Section 1,1-20
Transcript name: Giving and taking 10
Type of interaction: Conversational psychotherapy (10th session)
Interacting persons: K: Client (female), B: Therapist (male)
Audio recording: B; open
Transcription: R. Weingarten
Transcription system: Kallmeyer & Schütze

1, 1 K	_...mir gings-,die ganze Woche-, _also ich hab die Woche über
	How the whole week went for me? Then I have to think the week over
	_1 _2
2 B	
3 K	nich mehr-,so furchtbare Stimmungsschwankungen gehabt ne'
	Never before, never had such terrible mood swings
4 B	
5 K	_ so-..ging mir eher-,äh bescheiden. ..._(langsam) stimmungs-
	so granted, I have rather, um, tendency
	_3 _4
6 B	mhm
7 K	mäßig +_...ähm...ja,was jetz so akut grade war' gestern hat
	moderately +?... eh... yes, now what I felt so acutely yesterday has...
	_5
8 B	

9 K	mich mein Freund angerufen- _das war doch n ziemlicher
	my friend calling me? That was nevertheless a considerable
	_6
10 B	Mhm
11 K	Schlag vors Kontor wieder' _weil ich hatte das Gefühl-
	Shot across the kidneys again ' because I had the feeling
	_7
12 B	mhm _hat dich
	mhm you had
	_8
13 K	ich weiß es nich,hinter-
	I didn't know it, behind
14 B	irgendwie mitg-,mitgenommen schon.
	somehow it had happened already.
15 K	her ich hab mich selbst so-,einerseits ha-,hats mich nich
	A while ago, on the one hand hectar, hadn't
16 B	
17 K	so ausn Schuhen geschmissen wie ich dachte ne' so-_
	Knocked me out of my socks like it did' -?
18 B	_mhm,hattest
	Mhm, had
	_9
19 K	aber- ja'
	however
20 B	dir so- ,wirklich so vo- (k) schlimmer vorgestellt. denn
	because you had so, really made out so badly (k) .

The transcript shows *a complex experience thematization*, which also covers a report of a past situation. With this complex experience thematization, the client brings a past experience into the therapy situation, where it is then processed by therapist and client in specific ways.

In segment 1 [... I felt-, the whole week-], the client begins with the experience declarative formula "mir gings" (I felt). As segment 2 shows [well I didn''t have such terrible mood swings all week], a recapitulatory description of the experience is to be given. This is implemented in segment 2 by a negatively defined experience label "nicht mehr so furchtbare Stimmungsschwankungen" (not such terrible mood swings). In segment 3 [so-.. I felt more-, mhm moderately], experience is formulated positively, now again with an experience declarative formula and a short comparison. I characterize this as a short comparison, because in my opinion "bescheiden" (moderately) does not represent a conventional experience label. Segment 4 [(slowly) concerning mood], which is related to 3, shows and clarifies retroactively that this was meant as experience thematization. This elucidation may be connected with the fact that the suitability of "bescheiden" (moderately) for thematization of the experience appears questionable to K. Segment 5 involves the thematization of an experience by designation of an experience-relevant event: "gestern hat mich mein Freund angerufen" (yesterday my friend called me). From what follows, it becomes clear that this concerns the ex-friend. For all participants it is evident, and this also is supported by the introduction "was jetzt so akut grade war" (what just now was so acutely) that this event must have been connected with an intensive experience for K. The event is identified to thematize the concerning experience. Even if this experience had not been more exactly specified in the following utterances by means of other practices, B could have inferred it on the basis of emotion rules from the designated event. Segment 6 is then a description of the experience [that was still a considerable shot across the kidneys again]. Together with segment 5 and segment 7 [because I had the feeling-] K here supplies a complex experience thematization. The description of the experience in 6, which is carried out with a fixed figurative-metaphorical idiom, above all focuses the aspects of the intensity (considerable), the dynamics of the experience concerned (suddenness: "shot") and probably the type of experience.

It is notable that the client uses the figure of speech "shot across the kidneys." This German idiom is usually phrased "a shot to the kidneys." We can speculate that this is a confounding of two idiomatic phrases: "shot to the kidneys" and "shot across the bow." However these speculations transcend what should be demonstrated here. If something meaningful lies behind this slip of the tongue, it is of more concern to the therapist or analyst than to the linguist. In segment 7, the client sets up a reason for the intensity and dynamics of the experience described, which is initiated with an experience declarative formula "ich hatte das Gefühl" (I had the feeling). Now however in segment 8 B intervenes with a determining experience interpretation "hat dich irgendwie mitgenommen schon" (still stirred you anyhow). This is a typical communicative activity of therapists in client-centered therapy. There are several conceivable explanations for this intervention: The therapist can be of the opinion that the client has not thematized the experience clearly enough with the figurative description, or that it does not focus the relevant aspects of the experience. B focuses the consequences of the experience in his intervention. Last but not least, it is also conceivable that B only wants to paraphrase the description of the experience to ensure his understanding.

K understands the intervention as a focusing of possible consequences of the experience and rejects the imputed consequences: "einerseits hats mich nich so ausn Schuhen geschmissen wie ich dachte" (on the one hand it didn't knock me out of my socks in the way I thought it would). This figure of speech clearly focuses the possible consequences of the experience, whereby the whole utterance represents a statement regarding the unexpectedly small intensity of the experience. Again B intervenes with an utterance, which conceives the intensity verbal designative: "hattest dir *schlimmer* vorgestellt" (you thought it would be *worse*). This may throw a light on the intention of the first intervention retroactively. Under the circumstances, B would like to reflect back the figurative descriptions of the experience to K in the form of verbal labels. A resort to socially preformed and standardized verbal labels surely creates clarity for K with the interpretation of his emotional life. However, this use remains to be weighed against the plasticity and intensity of figurative descriptions of the experience.

The first transcription emphasized the manner in which the client brings her experience into the therapeutic conversation. In the next example, we focus on a typical activity of the therapist: the *experience focusing* already mentioned. In this example, it leads to a developed *experience exploration* of the client.

(II) Marriage problem 1, Transcript Section 23,35 - 24,18
Transcript name: Marriage problem 1
Type of interaction: Conversational psychotherapy (1st session)
Interacting persons: K: Client (male), B: Therapist (male)
Audio recording: B; open
Transcription: R. Weingarten
Transcription system: Kallmeyer & Schütze

23,35 K	daß da n Verhältnis is.
	that there is a relationship.
36 B	äh, wär das für <u>Sie</u> irgendwie (&) wenn
	Um, if that was so for you somehow (&) if
37 K	
38 B	Ihre Frau das denken würde oder sie denkt ja, + em-, was emp-
	Your wife would think that or she thinks yeah, + em, what fee-
39 K	... (leise) was emp-
	...(quietly) which fee-
40 B	finden Sie dabei daß Ihre Frau das denkt.
	Do you find that your wife thinks that.
24, 1 K	find ich dabei + ...ich fühl mich ja an und für sich ziemlich
	I think... I feel quite the same for myself and her
2 B	
3 K	unwohl dabei. ..(holt Luft)
	unwell with it. (draws a breath)
4 B	ähä, (leise) <u>was</u> is das dann fürn Gefühl +
	aha, (quietly) <u>what</u> kind of a feeling is that then?

5 K	wie soll ich das erklären,(sehr leise) es is n Gefühl- (7 Sek.
	Like I say, (very quiet) it is a feeling-(7 sec..
6 B	
7 K	(Pause) tja ich,+ also auf jeden Fall in mir dieses Unwohl-
	(Break) Yeah, I, in any case there is this sick feeling in me.
8 B	
9 K	sein. (h) äh is vorhanden äh
	its. (h) um is available um
10 B	ähä is da son Stück schlechtes Ge-
	Uh huh, you have a bad conscience about it.
11 K	ja,könnt ich (k) könnte man wohl sagen (sehr
	yes, can I (k) one could probably say (very much)
12 B	wissen dran',so daß
	know something about it, so that
13 K	(leise) das is sone Art- es is ne Art + schlech-
	(quietly) There is some type …it is a type + bad
14 B	daß man eigentlich sowas nicht tut,oder was
	that one actually does not do something like that, or which
15 K	tes Gewissen schon dabei. aber das Gros is glaub ich noch was
	bad conscience already. However I still believe most of it
16 B	ja
	yes
17 K	anderes. (8 Sek. Pause) ich mache mir ja auch-,äh *so* Gedanken
	other one. (8 sec. break) I also, um, have thoughts *like this*.
18 B	Ähä
	uh-huh

K takes the perspective of his wife in a problem description and sees the authorization for her suspicions that he has a relation with another woman. Thereupon B changes the perspective and focuses K's experience with the question: "was empfinden Sie dabei daß Ihre Frau das denkt" (what do you feel that your wife thinks that) (23,38-40). K repeats the question for himself and begins an experience exploration. He describes his experience with an experiential declarative formula "ich fühl mich" (I feel) (24,1) and the label "unwohl" (unwell) (24,3). Thereupon B requires a further specification "*was* is das dann fürn Gefühl" (*what* kind of a feeling is it) (24,4). B thematizes his difficulties during the description formulaically and repeats it (24,5-9). That is, he is not capable of a specification here. Now B continues with an designative projective experience interpretation: "is da son Stück schlechtes Gewissen dran" (you have a bad conscience about it) (24,10-12). B agrees with that, hesitating, but then determines: "aber das Gros is glaub ich noch was anderes" (however the essential I believe is something else) (24,15-17). This is not executed however, but after a break K changes to a new topic, with which he leaves the level of focusing on the experience, without B insisting.

CONCLUDING REMARKS

The attempt to understand emotionality from a decidedly interactive and social perspective is not only the prerequisite for making it accessible for a linguistic and conversation analytic handling but helps also to release emotions from a tendency to reify them as natural phenomena and "strange powers." They become recognizable as quantities that can be shaped and influenced communicatively and as resources in the process of interaction.

Likewise it becomes clear that they are not a purpose in themselves. Manifestation, interpretation, and processing of emotions always take place in larger frames for superordinate purposes. Emotions are manifested, interpreted, and processed in order to comfort someone, to solve problems, to carry out a conflict, to have fun together, to give therapy to a person, and there may be still other purposes. Communication processes related to feelings are embedded functionally in more global social practices. Apart from extensive analyses of emotions in different natural conversations, the explication of these frames belongs among the pending research tasks.

REFERENCES

Bamberg, M. (1997). Emotion talk(s): The role of perspective in the construction of emotions. In S. Niemeier, & R. Dirven (Eds.),*The language of emotions. Conceptualization, expression, and theoretical foundation* (pp. 209-225). Amsterdam: John Benjamins.

Battacchi, M. W., Suslow, T., &Renna, M. (1996). *Emotion und Sprache.* Frankfurt: Peter Lang.

Bliesener, T. (1982). Konfliktaustragung in einer schwierigen "therapeutischen Visite." In K. Köhle & H. Raspe (Eds.), Das Gespräch während der ärztlichen Visite (pp. 249-268). Munich: Urban & Schwarzenberg.

Bliesener, T., & Köhle, K. (1986). Die ärztliche Visite. Chance zum Gespräch. Opladen: Westdeautscher Verlag.

Bloch, C. (1996). Emotions and discourse. *Text, 16*, 323-341.

Brünner, G. (1987). Kommunikation in institutionellen Lehr-Lern-Prozessen. Tübingen, Germany: Narr.

Caffi, C., &Janney, R. W. (1994). Toward a pragmatics of emotive communication. *Journal of Pragmatics, 22*, 325-373.

Christmann, G. B., & Günthner, S. (1996). Sprache und Affekt. Die Inszenierung von Entrüstungen im Gespräch. *Deutsche Sprache, 24*, 1-33.

Cornelius, R. R. (1996). *The science of emotion. research and tradition in the psychology of emotions.* Upper Saddle River, NJ: Prentice Hall.

Coulter, J. (1979). *The social construction of mind.* London: Macmillan.

Danes, F. (1987). Cognition and emotion in discourse interaction: A preliminary survey of the field. In *Vorabdruck der Plenarvorträge*. XIV. Internationaler Linguistenkongreß unter der Schirmherrschaft des CIPL (pp. 272-291). Berlin.

Drescher, M. (1997). *Sprachliche Affektivität: Darstellung emotionaler Beteiligung am Beispiel von Gesprächen aus dem Französischen*. Bielefeld: Habilitationsschrift.

Ehlich, K. (1986). *Interjektionen*. Tübingen, Germany: Max Niemeyer.

Ehlich, K., & Rehbein, J. (1986). *Muster und Institution*. Tübingen, Germany: Narr.

Elias, N. (1981). *Über den Prozeß der Zivilisation*. Frankfurt: Suhrkamp.

Fiehler, R. (1986). Zur Konstitution und Prozessierung von Emotionen in der Interaktion. In W. Kallmeyer (Ed.), *Kommunikationstypologie. Jahrbuch 1985 des Instituts für deutsche Sprache* (pp. 280-325). Düsseldorf: Schwann.

Fiehler, R. (1990a). *Kommunikation und Emotion. Theoretische und empirische Untersuchungen zur Rolle von Emotionen in der verbalen Interaktion*. Berlin: De Gruyter.

Fiehler, R. (1990b). Erleben und Emotionalität als Problem der Arzt-Patienten-Interaktion. In K. Ehlich, A. Koerfer, A. Redder, & R. Weingarten (Eds.), *Medizinische und therapeutische Kommunikation. Diskursanalytische Untersuchungen* (pp. 41-65). Opladen: Westdeutscher Verlag.

Fiehler, R. (1990c). Emotionen und Konzeptualisierungen des Kommunikationsprozesses. *Grazer Linguistische Studien, 33/34*, 63-74.

Fries, N. (1991). Emotionen: Experimentalwissenschaftliche und linguistische Aspekte. *Sprache und Pragmatik, 23*, 32-70.

Fries, N. (1996). Grammatik und Emotionen. *Zeitschrift für Literaturwissenschaft und Linguistik, 101*, 37-69.

Gaus, E., & Köhle, K. (1982). Ängste des Patienten—Ängste des Arztes. Anmerkungen zur Konfliktaustragung in einer schwierigen Visite bei einem Todkranken. In K. Köhle & H. Raspe (Eds.), *Das Gespräch während der ärztlichen Visite* (pp. 269-286). München: Urban & Schwarzenberg.

Günthner, S. (1997). The contextualization of affect in reported dialogues. In S. Niemeier & R. Dirven (Eds.), *The language of emotions. Conceptualization, expression, and theoretical foundation* (pp. 247-275). Amsterdam/Philadelphia: John Benjamins.

Hartung, W. (1996). wir könn=n darüber ruhig weitersprechen bis mittags wenn wir wollen. Die Bearbeitung von Perspektiven-Divergenzen durch das Ausdrücken von Gereiztheit. In W. Kallmeyer (Ed.), *Gesprächsrhetorik. Rhetorische Verfahren im Gesprächsprozeß* (pp. 119-189). Tübingen, Germany: Narr.

Herrmann, T. (1987). *Gefühle und soziale Konventionen*. Arbeiten der Forschergruppe Sprache und Kognition am Lehrstuhl Psychologie III der Universität Mannheim. Bericht Nr. 40.

Hochschild, A. R. (1979). Emotion work, feeling rules, and social structure. *American Journal of Sociology, 85*, 551-575.

Kallmeyer, W. (1979a). "(expressif) eh ben dis donc, hein' pas bien'" - Zur Beschreibung von Exaltation als Interaktionsmodalität. In R. Kloepfer (Ed.), *Bildung und Ausbildung in der Romania* Bd. 1. (pp. 549-568). München: Fink.

Kallmeyer, W. (1979b). Kritische Momente. Zur Konversationsanalyse von Interaktions-störungen. In W. Frier & G. Labroisse (Eds.), *Grundfragen der Textwissenschaft* (pp. . 59-109). Amsterdam: Rodopi.

Käsermann, M.-L. (1995). *Emotion im Gespräch. Auslösung und Wirkung*. Bern: Huber.

Kövecses, Z. (1990). *Emotion concepts*. New York: Springer Verlag.

Konstantinidou, M. (1997). *Sprache und Gefühl*. Hamburg: Buske.

Lakoff, G., &Johnson, M. (1990). *Metaphors we live by*. Chicago: University of Chicago Press.

Lalouschek, J. (1993). "Irgendwie hat man ja doch bißl Angst." Zur Bewältigung von Emotion im psychosozialen ärztlichen Gespräch. In P. Löning & J. Rehbein (Eds.), *Arzt-Patienten-Kommunikation. Analysen zu interdisziplinären Problemen des medizinischen Diskurses* (pp. 177-190). Berlin: De Gruyter.

Löning, P. (1993). Psychische Betreuung als kommunikatives Problem: Elizitierte Schilderung des Befindens und 'ärztliches Zuhören' in der onkologischen Facharztpraxis. In P. Löning & J. Rehbein (Eds.), *Arzt-Patienten-Kommunikation. Analysen zu interdisziplinären Problemen des medizinischen Diskurses* (pp. 191-250). Berlin: De Gruyter

Lörcher, H. (1983). *Gesprächsanalytische Untersuchungen zur Arzt-Patienten-Kommunikation.* Tübingen, Germany: Niemeyer.

Marten-Cleef, S. (1991). *Gefühle ausdrücken. Die expressiven Sprechakte.* Göppingen.

Scherer, K. R., Summerfield, A. B., & Wallbott, H. G. (1983). Cross-national research on antecedents and components of emotion. A progress report. *Social Science Information, 22,* 355-385.

Schwitalla, J. (1991). Sozialstilistische Unterschiede beim Umgang mit dem "positiven Image". Beobachtungen an zwei Frauengruppen. In *Begegnung mit dem "Fremden."* Akten des VIII. Internationalen Germanisten-Kongresses Tokyo 1990. Bd. 4 (473-482). München: Judicium.

Schwitalla, J. (1996). Beziehungsdynamik. In W. Kallmeyer (Ed.), *Gesprächsrhetorik. Rhetorische Verfahren im Gesprächsprozeß* (pp. 279-349). Tübingen: Narr.

Schwitalla, J. (1997). *Gesprochenes Deutsch. Eine Einführung.* Berlin: Schmidt.

Selting, M. (1994). Emphatic speech style—with special focus on the prosodic signaling of heightened emotive involvement in conversation. *Journal of Pragmatics, 22,* 375-408.

Selting, M. (1995). *Prosodie im Gespräch. Aspekte einer interaktionalen Phonologie der Konversation.* Tübingen, Germany: Niemeyer.

Strauss, A, Fagerhaugh, S., Suczek, B., & Wiener, C. (1980). Gefühlsarbeit. Ein Beitrag zur Arbeits- und Berufspsychologie. *Kölner Zeitschrift für Soziologie und Sozialpsychologie, 32,* 629-651.

Thimm, C., & Kruse, L. (1993). The power-emotion relationship in discourse: spontaneous expression of emotions in asymmetric dialogue. *Journal of Language and Social Psychology, 12,* 81-102.

Tischer, B. (1993). *Die vokale Kommunikation von Gefühlen.* Weinheim: Beltz.

Vangelisti, A. L., Daly, J. A., & Rudnick, J. R. (1991). Making people feel guilty in conversations: techniques and correlates. *Human Communication Research. 18,* 3-39.

PART II

Figurative Language in Emotional Communication

– 5 –

Emotion Concepts: Social Constructionism and Cognitive Linguistics

Zoltán Kövecses
Eötvös Loránd University

One of the major figures in the social constructionist movement in the study of emotion is the philosopher Rom Harré. Because Harré's view, which he sometimes calls "emotionology," is an extremely influential one and because the cognitive linguistic view of emotion that I have been working on in the past 15 years bear certain important similarities, it makes sense to survey these similarities, as well as the differences, between his and my views. As will be seen, the cognitive linguistic view is in many ways sympathetic to Harré's proposals. The basic similarity between them is that in both theories language is seen as playing an important role in the study of the nature of emotion concepts. One of the differences between them appears to be in how these linguistic programs are carried out in this study. Emotionology is a heavily linguistic-semantic program, but its program cannot be fully carried out because emotionology does not have the appropriate kind of linguistics necessary for the analysis that the program sets out to accomplish. The major difference between the two approaches is that emotionology, and social constructionism in general, claims a high degree of cultural relativity at the expense of universality, whereas the cognitive view gives equal weight to both relativistic and universal factors in the conceptualization of emotion. In this chapter, I discuss in some detail the major similarities and differences between the two positions.

EMOTION VOCABULARIES IN DIFFERENT LANGUAGES

One of Harré's main ideas is that the different languages are characterized by different emotion terminologies. This chapter shows that in Harré's emotionology (a part of what he calls "discursive psychology") it is important that we investigate the complete terminology of emotions within a given language. The question arises as to what Harré means by an emotion terminology. In his papers (e.g., Harré, 1986, 1994), we find such emotion terms as *anxiety, joy, anger, sadness, boredom, embarrassment, jealousy.* These examples clearly show certain tendencies concerning this issue in Harré's thinking. First, by an emotion vocabulary he means those emotion words that are most commonly used by speakers of a language (in this case, English). Second, all the words just listed are of the kind that we could characterize as literal (rather than metaphorical). Third, the words on the list all indicate different emotions. In other words, the picture that Harré paints of emotion vocabularies suggests that this vocabulary is a collection of the most commonly used nonmetaphorical words denoting different emotions. For Harré (especially in his 1994 paper), the issue of what we mean by emotions means the "language games" that we can play with a few dozens of emotion words (on the basis of four criteria listed by him). This is important in emotionology in order to be able to show the differences between roughly similar emotions in any two languages-cultures (such as English, for instance, *anger* in English and its approximate Chinese counterpart *nu*)

However, Harré's 1994 paper is somewhat misleading concerning the program of emotionology. Harré clearly saw that within any given emotion, not only one but several "language games" can be played. That is, we should study the use of not just one emotion word within each domain but the use of several words in the same domain. This idea appears in another of Harré's writings (1986), in which he introduced "social constructionism." He wrote: "Instead of asking the question, 'What is anger?' we would do well to begin by asking, 'How is the word *anger*, and other expressions that cluster around it, actually used in this or that cultural milieu and type of episode?'" (p. 5).

What is important for us in this quote for the present purposes is the part where Harré talked about "*anger,* and other expressions that cluster around it." In other words, we are asked to imagine emotionology as an approach that investigates the use of not just one but several words even within the same emotion domain. However, this idea is not carried out by Harré or the other representatives of the social constructionist movement. They appear to be content with examining the use of a few key emotion words (such as *anger, fear*) in the way suggested by Harré. This practice, I believe, has certain negative consequences for the theory, to which I return later in the chapter.

Indeed, how many language games do we play in the case of an emotion? To put the same question more simply, how many linguistic expressions do we use in connection with our emotions? The number is much greater than what appears in the practice of constructionism. In the case of anger, there are at least 150 expressions available to speakers of English (Lakoff & Kövecses, 1987) and in the case of love the number is at least roughly 300 (Kövecses, 1988). Based on my studies of the English emotion lexicon (see, e.g., Kövecses, 1986, 1988, 1990, 1991a, 1991b), we can estimate the number of available language games for each basic emotion to be fairly high, in most cases over a hundred. We have no reason to believe that this number is any less in other languages (such as Chinese, for which see King, 1989, and Yu, 1995; or Hungarian, for which see Kövecses, 2000, Bokor, 1997). Obviously, the number of available language games in a language-culture depends heavily on the extent to which an emotion is viewed as foregrounded in the culture, that is, as "hypercognized" or "hypocognized," to use Levy's (1973) terms.

THE ROLE OF FIGURATIVE EXPRESSIONS

By "figurative expressions" I simply mean metaphors and metonymies. If we examine the several hundred linguistic expressions that are commonly used by native speakers of, say, English, to talk about the emotions, we find that most of these are figurative, that is, metaphoric or metonymic in nature. Speakers of English say that people *boil* with anger, *tremble like a leaf, burn* with desire, *give vent* to their feelings, *hold back* their emotions, are *overwhelmed* by joy, are *hit* by somebody's death, can be *puffed up* or *swelled* with pride, can be *hot* with lust, can be *sustained* by hope, and many others. The role of these and many other similar linguistic expressions in Harré's work on emotion is unclear, and the other representatives of the social constructionist view attribute no (or very little) importance to them.

By ignoring figurative language, however, social constructionism leaves unexplored one of the major factors in (either folk or expert) theory making. Metaphor has the power to create reality for us; it is the major way in which the human cognitive system produces nonphysical reality, that is, the social, political, psychological, emotional, etc. worlds (see Kövecses, 1999, 2000, in press). Many, if not all, expert and nonexpert theories in these general domains are based on what Lakoff and Johnson (1980) called "conceptual metaphor" and are reflected linguistically in metaphoric expressions. Some well-known and studied examples from domains outside emotion include the *superstructure* of the Marxist theory of society, which relies on the conceptual metaphor of SOCIETY IS A BUILDING (Rigotti, 1995), the *sending* and *receiving* of messages, which relies on the conceptual metaphor complex MEANINGS ARE OBJECTS, LINGUISTIC EXPRESSIONS ARE CONTAINERS (FOR THESE OBJECTS), and COMMUNICATION IS SENDING OBJECTS (Reddy, 1993 and Lakoff, 1993), and the conception of the human mind as a computer, which relies on the metaphor THE MIND IS A COMPUTER (Sternberg, 1990).

In one dominant view of cognitive science and cognitive linguistics (as represented in the work of Lakoff, 1987, Johnson, 1987, and Gibbs, 1994), metaphor is not simply a linguistic expression with a non-literal meaning. Metaphors work at a conceptual level and the linguistic expressions merely reflect certain conceptual structures (such as THE MIND IS A COMPUTER). In several studies I have suggested that the emotions are "constructed" by means of such conceptual devices, most prominent among these being the conceptual metaphor EMOTION IS FORCE. (On the controversy between Quinn and Lakoff and myself concerning this issue, see Quinn, 1991; Strauss & Quinn, 1997, Kövecses, 1995b, 1999.) There are, of course, universal and culture-specific aspects to this conceptualization, which I analyze elsewhere (see Kövecses, 2000).

Although Harré's 1994 paper does not mention metaphor and its role, in the other paper mentioned earlier (Harré , 1986) he does make reference to it, again in a programmatic form:

> ... we do say that someone is *puffed up* or *swollen with pride*, too. These metaphors may perhaps be traced to an element of the ridiculous in an exaggerated or excessive display. The matter deserves more research. The same could be said for hope, which also benefits from a cluster of characteristic metaphors, such as *surging*, *springing* and the like (p. 9).

That is, Harré is well aware that metaphor is important in a constructivist view of emotion and deserves further research, but he and other constructivists do not explicate its relevance, nor do they demonstrate it through detailed case studies. In this respect, then, the "real constructivists" are those cognitive linguists, who, following Lakoff and Johnson's (1980) theory of metaphor, view emotion concepts as being largely (though not completely) constituted by metaphor. This is the theory on the basis of which emotion concepts can be claimed to be social-cognitive constructions.

I do not wish to go into the technical details of how emotion concepts emerge as conceptual structures constituted by metaphor (but see Kövecses, 1991b, in press). Suffice it to say that, as was already mentioned, the generic metaphor speakers of English (and other languages as well) most heavily rely on in understanding what the emotions are is EMOTION IS A FORCE. This force can be human (e.g., in the specific metaphor EMOTION IS AN OPPONENT), animal (e.g., in EMOTION IS A WILD ANIMAL), physical (e.g., in EMOTION IS A MAGNETIC/GRAVITATIONAL FORCE), natural (e.g., EMOTION IS A FLOOD/STORM), or a force influencing human perception or thought (e.g., in EMOTION IS INSANITY/RAPTURE). This particular conceptualization goes with a certain logic. For ordinary people emotions are FORCES that emerge independently of a rational and conscious self as a result of certain causes, and that, in most cases, have to be kept under control. In other words, the role and significance of metaphor in emotion is that it creates a certain model of emotion. This aspect of the study of emotion is completely missing from, or is present only programmatically in, the view of emotion eminently represented by Harré.

COGNITIVE MODELS OF EMOTION

When we say that the metaphors produce a certain model of emotion (that is, provide a certain conception of it), what I have in mind is a cognitive structure that is variously called "schema," "script," "cultural model," "cognitive model," "idealized cognitive model," and the like, in psychology, anthropology, and linguistics (for a recent introduction to this concept, see Strauss & Quinn, 1997). The language-based cultural model of emotion in English comprises several stages that unfold in time. This generic cognitive model can be given as follows (based on Kövecses, 1990, chap. 11). In the model only the major aspects of the concept are given that are produced by the generic metaphor EMOTION IS A FORCE, thus ignoring several additional aspects of the concept.

0. Neutral emotional state
   The subject (S) is emotionally calm.
1. Cause
   Something happens to S.
   The event exerts a sudden and strong impact on S
   Emotion (E) comes into existence
   S is passive with regard to this.
2. Emotion exists
   Emotion acts as a force on S.
   Part of emotion is a desire to cause S to perform an action.
   S knows that the act is socially dangerous and/or unacceptable to do.
   The action, if performed, can satisfy the desire involved in emotion.
   The intensity of emotion is high; it is near the limit that S can control
3. Control
   S knows that he is under obligation to resist the desire and not to perform the action.
   S applies a counterforce to prevent the action from happening.
   However, the intensity of emotion as a force increases over the limit that S can control.
4. Loss of control
   S is now unable to control the force acting on him/her.
   The force causes S to perform the action.
5. Action
   S performs the action.
   S is not responsible for the action, because he/she only obeys a stronger force.
   The desire in emotion is now satisfied.
   Emotion ceases to exist.
0. Neutral emotional state
S is calm again.

This is the model of emotion that a language-based study yields, but it is not just a model that inheres in language. It also exists in people's heads. Parrott

(1995) demonstrated the psychological reality of the model with sociopsychological experiments. As I have already emphasized, the model represents a certain folk theory of emotion. And it is increasingly certain that it does not represent something that really happens when people experience emotion, despite the fact that it exists in people's heads. The scientific theory of emotion that is gaining more and more acceptance as a result of neurobiological experiments is closer to the Jamesian view of emotion, in which the "response" precedes the experience of emotion itself (see Le Doux, 1996).

I believe that given this idealized cognitive model of emotion we can understand better how metaphors are capable of producing a certain conception of emotion. Both emotion and the cause of emotion are metaphorically viewed as concrete forces. The cause-as-force produces the emotion and the emotion-as-force produces a response. The rational self is also viewed as a forceful agent that attempts to control emotion but, in the prototypical case, eventually gives in to its stronger force. This yields a generic-level structure to the concept of emotion that can be given as: "cause→emotion→response." Without conceiving of emotion metaphorically as concrete forces, it would be difficult to see how this particular generic level model of emotion could have emerged (see Kövecses 1995a, 1995c). Given this inherently metaphorically-structured concept, Quinn's (1991) claim that the cultural model of anger is literal does not seem to be right. (On the controversy concerning whether abstract concepts are literally or metaphorically constituted, see Kövecses 1995c, 1999.)

The model just described provides the "prototype" of emotion—in the sense of Rosch (e.g., 1978). Prototypical emotions include anger, fear, joy, and sadness. All of these can be characterized in roughly the terms of the model as described. Needless to say, this is just one of the many commonsense models of emotion that people have. What gives it privileged status is the fact that it is a central one from which all kinds of deviations are possible. These "deviations" represent further, less prototypical cases. Less prototypical cases include situations where, in "weaker" emotions, the issue of control does not even arise or where, at the end of an intense emotional episode, the self does not calm down but remains "emotional." There are many such additional nonprototypical cases. Specific emotions can also be represented in terms of prototypical models. Anger, fear, joy, love, and so on also exist in many forms, which characterize various deviations from their respective prototype, or best example (see Lakoff & Kövecses, 1987, for anger; Kövecses, 1990, for fear; Kövecses, 1988 and 1991a, for love; and Kövecses, 1991b, for joy and happiness). Such nonprototypical cases are often given linguistic manifestations that are different from that denoting the prototype. For example, *giving vent to* one's emotion describes a controlled way of expressing one's feelings (as opposed to losing control over it against one's will), and is thus a nonprototypical form of emotion; *indignation* is a nonprototypical form of anger in which a wrongdoer does harm not to the self (the subject of anger) but typically to a third party. Significantly, many of these nonprototypical cases are metaphorical in nature, as the example of *give vent to* one's emotion indicates (as it is based on the metaphor according to which the BODY IS A CONTAINER and EMOTION IS A FLUID UNDER PRESSURE IN IT). What this shows is that we play many "language games" in connection with both the generic concept of emotion and the specific emotion concepts. In addition, may of these language games are metaphorical. Social constructionists lose sight of the fact that both the prototypical and nonprototypical cases of emotion may be constituted by figurative language. This may seem like a radical idea, but it can be taken even farther. It can be claimed that there is nothing in the conceptualization of emotions that is not figurative. Györi (1998) showed that emotion words that we take to be literal (nonfigurative) today are etymologically all figurative. *Anger, grief, happy* (English), *rad, gore, ljubov* (Russian), *Hass, Zorn, Strach* (German), and *düh, méreg, szeret,* and *szomorú* (Hungarian) are common emotion terms in these languages and are based on still-active conceptual metaphors and metonymies.

If we examine the content of the idealized cognitive models associated with *emotion* or other emotion concepts, we find that they greatly overlap with Harré's rules of emotion. According to Harré (1994), in the course of the appropriate use of emotion words in different cultures people observe certain "local rules." The rules are of four kinds, "classified by reference to what is criterial for their correct usage": (a) "appropriate bodily feelings," (b) "distinctive bodily displays," (c) "cognitive judgments," and (d) "moral judgments" and the "social acts" corresponding to them (p. 7).

Harré's four kinds of rules neatly match the different aspects of the prototypical folk model of emotion given previously. Physical sensations and bodily manifestations can be found in stage two of the cognitive model, where the subject of emotion produces certain physiological and behavioral responses to a certain event; cognitive judgments occur in stage one, where certain events are judged in certain ways, as a result of which the subject feels he or she is in a particular emotional state; moral judgments can be found in stages one and three, where the subject judges the event (stage one) and the emotion itself (more appropriately, the need to control it) according to the local moral code; and, finally, social actions occur in stage five, where the subject performs one or more social actions appropriate to the emotion. Given this overlap, it can be suggested that the rules of emotionology and the cognitive models of emotion(s) as described have largely the same content. It is important to see that both cognitive models of emotion and Harré's rules of emotionology characterize everyday, or folk, understandings of emotion—not scientific, or expert, ones. We can assume that, as the rules of emotionology emerge from an investigation of everyday language use, emotionology does not distinguish between the folk and the expert views of emotion (but see Kövecses, 1994, 2000, chap. 7, for the relationship between the two in cognitive linguistics).

RELATIVITY VERSUS UNIVERSALITY

One of the major ideas of emotionology and, more generally, of social constructionism, is that the "language games" played in different languages-cultures are very much unlike each other even in the case of roughly corresponding words. That is, as Harré repeatedly stressed in his 1986 paper, emotions are characterized by different emotion vocabularies in different cultures and, what's more, even the emotions themselves differ from culture to culture (Harré, 1986, p. 10).

Two questions arise: (a) Which emotion-related conceptual metaphors are universal (or near-universal) and which ones are language-culture specific? (b) Can we *predict* which emotion-related conceptual metaphors are universal? In other words, do we have any basis for predicting which of these are language-culture specific and which are not?

Cognitive linguistics is just beginning to pay attention to this issue, although it may well be that questions of universality are some of the most interesting and significant ones in cognitive science. The first question can be settled in an empirical way. We have to study as many languages as possible and check whether a given emotion-related conceptual metaphor that is present in *any one* language-culture can be found in other languages-cultures. This is no small task, as there are several dozens of emotion-related metaphors, for example, in English and there are thousands of other languages-cultures around the world. We can answer the second question only if we have reliable empirical evidence of the universality (or at least near-universality) of at least one emotion-related metaphor. In this case, we can begin to make hypotheses concerning the issue of why certain conceptual metaphors are universal (or near-universal).

Following my own work with Lakoff (Lakoff & Kövecses, 1987) on anger, several scholars have looked at languages other than English. This work has given me occasion to compare emotion-related conceptual metaphors in several radically different languages and cultures (see Kövecses, 1995a, 1995b, 1995c, 2000). The languages that were investigated in a detailed way with this goal in mind include English, Hungarian, Japanese, and Chinese. In addition, we have some data from Wolof, an African language spoken in Senegal and Gambia, and some observations about Tahitian culture. The conceptual metaphor related to anger that can be found in all of these languages is ANGER IS PRESSURE IN A CONTAINER, a kind of FORCE metaphor. In what follows in this section, I illustrate this metaphor with only a few examples.

In English, we find a special case of the metaphor: ANGER IS A HOT FLUID IN A CONTAINER. The hot fluid exerts pressure on the walls of the container (that is, the human body). Lakoff and Kövecses (1987) offered these metaphorical linguistic expressions for the HOT FLUID conceptual metaphor in English:

> He was *boiling.*
> Sam *exploded.*
> She is *seething.*
> I was *fuming* for hours.
> He was *pissed* off.
> After she *let off some steam*, she felt better.

As Matsuki (1995) showed, a similar conceptual metaphor is found in Japanese as well. In Japanese, the equivalent of anger is *ikari.* (Later, I provide the Japanese, Chinese, Hungarian, and Wolof examples as transcribed into English by the authors whose work I quote.) Here are some Japanese examples for the PRESSURIZED CONTAINER metaphor:

> Ikari ga karada no naka de tagiru.
> anger seethes inside the body
>
> Ikari ga hara no soko wo guragura saseru.
> anger boils the bottom of stomach

The nearest Chinese counterpart of *anger* is *nu* (see King, 1989, and Yu, 1995). As King and Yu pointed out, Chinese also has the PRESSURIZED CONTAINER metaphor. This can be demonstrated with such examples as the following:

> qi man xiong tang
> qi full breast
> to have one's breast full of qi
>
> qi yong ru shan
> qi well up like mountain
> one's qi wells up like a mountain
>
> bie yi duzi qi
> hold back one stomach qi
> to hold back a stomach full of qi
>
> bu shi pi qi fa zuo
> NEGATIVE make spleen qi start make
> to keep in one's spleen qi

The word for *anger* is *düh* in Hungarian. My students and I have studied this concept and found that the same conceptual metaphor is present in Hungarian as well (Bokor, 1997). Some examples include:

> Forrt benne a düh. [boiled in-him the anger]
> Anger was boiling inside him.

Fortyog a dühtöl. [seethed the anger-with]
He is seething with anger.

The HOT FLUID metaphor is also found in Wolof. Munro (1991) observed that the Wolof word *bax*, which has the primary meaning "to boil," also possesses the meaning "to be very angry."

We have some evidence of this conceptual metaphor in Tahitian. According to Levy, as quoted in Solomon (1984), "The Tahitians say that an angry man is like a bottle. When he gets filled up he will begin to spill over" (p. 238).

As these examples show, *anger* and its counterparts in several different languages-cultures are conceptualized by means of remarkably similar conceptual metaphors. (Needless to say, there are several interesting and important differences in this conceptualization, but I do not discuss them here. See Kövecses, 2000). We can then tentatively suggest that the conceptual metaphor ANGER IS PRESSURE IN A CONTAINER is a near-universal metaphor. How is this possible? Several answers suggest themselves. First, we can say that the great degree of similarity among the metaphors used in these languages-cultures is simply an accident. For some reason, members of these cultures decided to use this particular conceptual metaphor. Second, one can claim that the PRESSURIZED CONTAINER metaphor emerged in one culture and then somehow it was transmitted to the others. The radically different cultures simply "borrowed" this way of conceptualizing anger from an original source language. Third, it can be suggested that people in these different cultures possess certain attributes that make them conceptualize anger in the same way. The shared attributes may be certain physiological features of the human body during the experience of anger. Although I consider the first two options as possibly playing some role, I believe that it is the third possibility that is most convincing.

More specifically, I claim that the human body and its physiological processes in anger are independent of culture, that is, they are universal. Thus, it is the similarity of the body and that of its physiological processes in anger that may motivate the emergence of the (roughly) same metaphorical conceptualization. Do we have any linguistic evidence to support this claim? It seems that we do. It is shown in the metonymies (that is, "stand for" relationships) associated with anger in the different languages-cultures. Metonymies in the domain of emotion describe physiological and behavioral responses in emotional states (see Kövecses, 1986, 1990). By making reference to one's physiological and/or behavioral responses, one can talk about emotion. (In other words, there is a "stand-for" relationship between response and emotion.) Which physiological and/or behavioral responses are used to talk about anger in the languages-cultures we have some evidence for? Let us survey these.

INCREASE IN BODY TEMPERATURE

English (examples from Lakoff & Kövecses, 1987):

> Don't get *hot under the collar.*
> Billy's a *hothead.*
> They were having a *heated* argument.
> When the cop gave her a ticket, she got all *hot* and bothered.

Chinese (Yu, 1995):

> Wo qi de lian-shang huo-lala de.
> I gas fire-hot

Japanese (examples from Noriko Ikegami & Kyoko Okabe):

> (Watashi-no) atama-ga katto atsuku-natta.
> my head get hot

> Karera-wa atsui giron-o tatakwasete-ita
> they heated argument were having

> atama o hiyashita hoo ga ii
> head cool should

Hungarian:

> forrófejü
> hotheaded

> felhevült vita
> heated debate

> *Lehül* egy kicsit.
> downcool a little

Wolof (examples from Munro [1991]):

> tang [to be hot]
> to be bad-tempered

> Tangal na sama xol. [he heated my heart]
> He upset me, made me angry.

Tahitian (Levy, 1973, does not contain data relating to body heat).

INTERNAL PRESSURE

English (examples from Lakoff & Kövecses ,1987):

> Don't *get a hernia*!
> When I found out, I almost *burst a blood vessel.*
> He almost *had a hemorrhage.*

Chinese (examples from King, 1989):

> qi de naomen chong xue

qi DE brain full blood

qi po du pi
break stomach skin

fei dou qi zha le
lungs all explode LE

Japanese (examples from Noriko Ikegami & Kyoko Okabe):

kare no okage de ketsuatsu ga agarippanashi da [he due to blood
pressure to keep going up]
he due to blood pressure to keep going up

sonna ni ikiri tattcha ketsuatsu ga agaru yo [like that get angry blood
pressure to go up]
like that get angry blood pressure to go up

Hungarian:

agyvérzést kap [cerebral-hemorrhage gets]
will have a hemorrhage

felmegy benne a pumpa [up-goes in-him the pump]
pressure rises in him

felment a vérnyomása [up-went the blood pressure-his]
His blood pressure went up.

Tahitian: (no data).
Wolof (no data).

REDNESS IN THE FACE AND NECK AREA
English:

She was *scarlet* with rage.
He *got red* with anger.
He *was flushed* with anger.

Chinese (examples from King, 1989):

ta lian quan hong le yanjing mao huo lai
he face all red LE eyes emit fire come

qi de lian dou zi le
qi face all purple

Japanese (examples from Noriko Ikegami & Kyoko Okabe):

kare wa makka ni natte okotta
he red to be get angry

makka ni natte okoru
red become get angry

kare wa ikari-de akaku-natta

he with anger got red

Hungarian:

Vörös lett a feje. [red became the head-his]
His head turned red.

Tahitian (no data).
Wolof (no data).

In my view it is these physiological responses coded as metonymies into a variety of languages that may have led to the similar conceptualization of anger and its counterparts in different cultures. We called this conceptualization the PRESSURIZED CONTAINER conceptual metaphor. Another part of the motivation for this may be that these cultures, and possibly others as well, conceive of the human body as a container, in which there is some hot fluid (e.g., the blood) that can exert pressure on the container. This physical pressure corresponds metaphorically to the force that may lead to a loss of control and which forces the angry person to perform certain (aggressive) actions. (There are many additional complications that I leave out of this account, but see Kövecses , 2000.)

It is crucially important for the cognitive linguistic view to ask whether the physiological processes in anger that were identified in language earlier are merely folk theoretical notions or they can be established objectively, that is, they are real. Levenson and his colleagues (1992) showed that Americans and members of the Minangkabau tribe living in West Sumatra produce the same physiological responses when they are angry: Among other things, their body temperature increases and their blood pressure rises. Levenson and Ekman (Ekman, Levenson, & Friesen, 1983: Levenson, Ekman, Heider, & Friesen, 1992) provided further evidence that metaphorical and metonymic conceptualization is based on universal human experiences, including, importantly, physiological ones—especially in the realm of the emotions.

These results point to the conclusion that emotionology and social constructionism go too far in claiming linguistic and cultural relativity in the domain of emotion. As we have seen, a large part of emotional conceptualization, due to universal physiology, appears to be universal.

CONCLUSIONS

It appears then that, contrary to the views of social constructionism, the conceptualization of emotions is, to some extent at least, universal (or near-universal). Obviously, this idea leaves room for the complementary view that several additional aspects of the emotions and the "language games" we can play with emotion terms can be greatly different in different languages and cultures (see Kövecses, 1995a, 1995b, 2000). The cognitive linguistic approach agrees with emotionology and social constructionism that emotion concepts are linguistically-culturally different, but disagrees with their radical relativity. The view that I find convincing is that, at least in the case of what are called "basic emotions," emotion concepts possess a solid bodily-physiological basis and that this basis leads to a certain degree of (near-)universality in the conceptualization of emotions.

In my view, the constructivist potential of emotionology is not, and cannot, be realized because it does not take seriously the "world making" potential of metaphor. Although in his program Harré pays some attention to and sees a part for metaphor in constituting the emotional world, this program is only realized in cognitive linguistics. In this sense, cognitive linguistics can be thought of as the "most fully accomplished form" of constructionism (but leaving behind its radical relativity).

This weakness of emotionology and constructivism largely follows from the fact that they confine the study of emotion to the analysis of a few commonly used, nonmetaphorical emotion words, instead of paying attention to the large number of words and expressions related to particular emotions and their richness and complexity that can be found in different languages of the world. Emotionology and constructionism cannot realize their own linguistic-semantic program that they share with cognitive linguistics. The cognitive linguistic view of emotions is capable of integrating a methodologically sound analysis of the linguistic richness and complexity of emotion language in a particular culture with social-cultural variation, as well as with universality that arises from the physiology of the human body.

REFERENCES

Bokor, Z. (1997). Body-based constructionism in the conceptualization of anger. *C.L.E.A.R. Series, No. 18,* publications of the Seminar für Englische Sprache und Kultur, Universitat Hamburg and the Department of American Studies, Eötvös Loránd University, Budapest.

Ekman, P., Levenson, R W., & Friesen, W. V. (1983). Autonomic nervous system activity distinguishes among emotions. *Science, 221,* 1208-1210.

Gibbs, R. (1994): *The Poetics of Mind.* Cambridge: Cambridge University Press.

Györi, G. (1998). Cultural variation in the conceptualization of emotions: A historical study. In A. Athanasiadou & E. Tabakowska (Eds.), *Speaking of emotions: Conceptualization and expression* (pp. 99-124). Amsterdam: John Benjamins.

Harré, R. (1986). An outline of the social constructionist viewpoint. In R. Harré (Ed.), *The social construction of emotion* (pp. 2-14). Oxford: Basil Blackwell.

Harré, R. (1994, May). *Emotion and memory: The second cognitive revolution.* Paper presented at the Collegium Budapest, Budapest, Hungary.

Johnson, M. (1987). *The Body in the Mind.* Chicago: Chicago University Press.

King, B. (1989). *The conceptual structure of emotional experience in Chinese.* Unpublished doctoral dissertation, Ohio State University.

Kövecses, Z. (1986). *Metaphors of anger, pride, and love.* Amsterdam: John Benjamins.

Kövecses, Z. (1988). *The language of love.* Lewisburg: Bucknell University Press.

Kövecses, Z. (1990). *Emotion concepts.* New York: Springer-Verlag.

Kövecses, Z. (1991a). A linguist's quest for love. *Journal of Social and Personal Relationships 8,* 77-97.

Kövecses, Z. (1991b). Happiness: A definitional effort. *Metaphor and Symbolic Activity, 6,* 29-46.

Kövecses, Z. (1994). Ordinary language, common sense, and expert theories in the domain of emotion. In J. Siegfried (Ed.)., *The status of common sense in psychology* (pp. 77-97). Norwood, NJ: Ablex.

Kövecses, Z. (1995a). The CONTAINER metaphor of anger in English, Chinese, Japanese and Hungarian. In Z. Radman (Ed.), *From a metaphorical point of view* (pp. 117-145). Berlin-New York: de Gruyter.

Kövecses, Z. (1995b). Anger: Its language, conceptualization, and physiology in the light of cross-cultural evidence. In J. R. Taylor & Robert E. MacLaury (Eds.), *Language and the cognitive construal of the world* (pp. 181-196). Berlin and New York: de Gruyter.

Kövecses, Z. (1995c). Metaphor and the folk understanding of anger. In J. A. Russell, J.-M. Fernández-Dols, A. S. R. Manstead & J. C. Wellenkamp (Eds.), *Everyday conceptions of emotion* (pp. 49-71). Dordrecht: Kluwer Academic Publishers.

Kövecses, Z. (1999). Metaphor: Does it constitute or reflect cultural models? In R Gibbs & G. Steen (Eds.), *Metaphor in cognitive linguistics* (pp. 167-188). Amsterdam: Benjamins.

Kövecses, Z. (2000). *Metaphor and emotion.* Cambridge: Cambridge University Press.

Kövecses, Z. (in press). *Metaphor: A practical introduction.* New York: Oxford University Press.

Lakoff, G. (1987). *Women, fire, and dangerous things.* Chicago: The University of Chicago Press.

Lakoff, G. (1993). The contemporary theory of metaphor. In A. Ortony (Ed.), *Metaphor and thought.* (2nd ed., pp. 202-251). Cambridge University Press.

Lakoff, G., & Johnson, M. (1980). *Metaphors we live by.* Chicago: The University of Chicago Press.

Lakoff, G., & Kövecses, Z. (1987). The cognitive model of anger inherent in American English. In D. Holland & N. Quinn (Eds.), *Cultural models in language and thought* (pp. 195-221). Cambrdige: Cambridge University Press.

Le Doux, J. (1996). *The emotional brain.* New York: Simon and Schuster.

Levenson, R.W., Ekman, P., Heider, K., & Friesen, W. V. (1992). Emotion and autonomic nervous system activity in the Minangkabau of West Sumatra. *Journal of Personality and Social Psychology, 62,* 972-988.

Levy, R. I. (1973). *Tahitians: Mind and experience in the Society Islands.* Chicago: University of Chicago Press.

Matsuki, K. (1995). Metaphors of anger in Japanese. In J. R. Taylor & R. E. MacLaury (Eds.), *Language and the cognitive construal of the world* (pp. 137-151). Berlin and New York: de Gruyter.

Munro, P. (1991). *ANGER IS HEAT: Some data for a crosslinguistic survey.* Unpublished manuscript, University of California at Los Angeles.

Parrott, G. W. (1995). The heart and the head. Everyday conceptions of being emotional. In J. A. Russell, J.-M. Fernández-Dols, A. S. R. Manstead & J. C. Wellenkamp (Eds.), *Everyday conceptions of emotion.* (pp. 73-84). Dordrecht, Netherlands: Kluwer.

Quinn, N. (1991). The cultural basis of metaphor. In J. Fernandez (Ed.), *Beyond metaphor* (pp. 56-93). Stanford, CA: Stanford University Press.

Reddy, M. (1993[1979]). The "conduit" metaphor. In A. Ortony (Ed.), *Metaphor and thought* (2nd ed., pp. 164-201). Cambrdige: Cambridge University Press.

Rigotti, F. (1995). The house as metaphor. In Z. Radman (Ed.), *From a metaphorical point of view* (pp. 419-445). Berlin-New York: de Gruyter.

Rosch, E. (1978). *Human categorization*. Hillsdale, NJ: Lawrence Erlbaum Associates.

Solomon, R. (1984). Getting angry: the Jamesian theory of emotion in anthropology. In R. A. Shweder & R. A. LeVine (Eds.), *Culture theory* (pp. 238-254). Cambridge University Press.

Sternberg, R. (1990) *Metaphors of Mind*. Cambridge University Press.

Strauss, C., & Quinn, N. (1997). *A cognitive theory of cultural meaning*. Cambridge: Cambridge University Press.

Yu, N. (1995). Metaphorical expressions of anger and happiness in English and Chinese. *Metaphor and Symbolic Activity, 10*, 9-92.

– 6 –

What's Special About Figurative Language in Emotional Communication?

Raymond W. Gibbs, Jr., John S. Leggitt, Elizabeth A. Turner
University of California at Santa Cruz

People often employ figurative language to verbally express their emotions. Ask someone to talk about an important part of his or her life, and metaphor, metonymy, irony, and other tropes will pervade their narrative. This chapter examines the role that figurative language plays in emotional communication. We argue that figurative language is especially useful for expressing the nuances of emotion, and for evoking particular emotional reactions in others, because it tightly reflects people's figurative conceptualizations of their emotional experiences. Figurative language is also special for the power it affords speakers to evoke particular emotions in others. We begin this chapter by describing speaker-listener dynamics in emotional communication. Next, we present several sections on the importance of figurative language in talking about and evoking emotional experiences. Following this, we describe two sets of empirical studies that looked at how listeners emotionally respond to metaphor and irony. We specifically show how emotional responses to metaphor and irony differ from those to literal language. Moreover, listeners draw different assumptions about what speakers intend to communicate when using figurative language compared to when they speak literally. Finally, we offer some challenges for future work on figurative language, emotion, and communication.

THE DYNAMICS OF EMOTIONAL COMMUNICATION

Understanding the role of figurative language in emotional communication requires some recognition of the interpersonal dynamics of conversational interaction. People speak figuratively for reasons of politeness, to avoid responsibility for the import of what is communicated, to express ideas that are difficult to communicate using literal language, and to express thoughts in a compact and vivid manner (Ortony, 1975). Consider the following narrative from a married man, Sam, where he discusses his wife's infertility and its effect on their lives (Becker, 1997):

> It (infertility) became a black hole for both of us. I was so happy when I was getting married, and life for me, was consistently getting better. And she was continuously depressed. Everything was meaningless because she couldn't have a baby. And so it was a tremendous black hole, it was a real bummer. I mean, in the broadest, deepest sense of the term. It was very upsetting to me because it was like no matter what... it seemed like every time... I was, like, taking off and feeling good, and she was dragging me down. And it wasn't that she was dragging me down, but she was dragging herself down. It was contrary to the overwhelming evidence of our lives, and it was very disturbing, to the point where I said to her that it was not tolerable anymore. And she went to a psychiatrist. She wanted to improve because she couldn't get out of it. I mean, she could not get out of it, and the Prozac (that was prescribed), made it worse. But it broke the cycle of depression, and she got out of it! But that was terrible. It ruined our sex life. It was just like everything was going down the black hole....The notion of the black hole is that it's this magnet—this negative magnet in space through which all matter is irretrievably drawn—that was the image that I had of it. It was just sucking everything down out of our lives. Down this negative hole. It was bad. (pp. 66-67)

Sam describes the black hole (the infertility) as a terrible constricting force that prevents him and his wife from experiencing the pleasures and responsibilities of parenthood. This black hole sucks significant, much desired elements out of their relationship, and as a metaphor, reflects a dismal image of their deteriorating marriage, one that both partners seem to have little control over. The comparison of the experience of infertility in marriage to being in a black hole is just one of several illustrations in this passage of how metaphor, in this case from the world of science, enables ordinary people to conceptualize, and verbally talk about, their emotional experiences. For Sam, metaphor expresses the outrage, frustration, and despair he feels over his recent situation. Imagine how a reader would think of Sam's emotional experience if he were to talk of his wife's infertility, and its effect on their marriage, using nonmetaphorical language. What if Sam only used the terms *outrage, anger, frustration*, or *despair*. For those of us who know even a little bit about black holes, Sam's expression *it (the infertility) became a black hole for both of us* provides an immediacy, a concrete vividness, to what he must have felt that literal language alone simply can't provide.

Reading Sam's narrative not only offers insights into his complex emotional state of mind, but also evokes in us emotions about Sam, ourselves, and perhaps our relationship with Sam (for those people who know Sam or listen to his narrative in person). Among our emotional reactions when first reading Sam's narrative were sympathy/compassion for Sam and his wife, a sharing of their deep frustration over the infertility problem, positive affect toward Sam for his sensitivity toward his wife, but also frustration at Sam's self-centeredness, and the couple's apparent inability to act more forcefully in trying to deal with their problem, a feeling that slightly tempers whatever sadness we feel over their situation. Regardless of the range of emotions Sam's narrative has evoked, we remain impressed by the aptness of his main *black hole* metaphor to capture what he must be experiencing emotionally and by how this, and other, metaphors in the passage work on us to create some degree of emotional solidarity between Sam and readers of his narrative.

We can speculate about what Sam's possible communicative intentions might be in describing his emotions in the metaphorical ways that he does. In fact, our reading of Sam's narrative immediately raises the question of what Sam was hoping to accomplish by talking about his emotional experiences in the way he did. Conversational analysts often assume that the expression and understanding of speaker meaning depends on, and is limited by, the recognition of communicative intentions (Gibbs, 1998, in press). Intentions are psychological states, and most scholars assume that the content of an intention must be mentally represented. In particular, a speaker or writer must have in mind a representation of the set of assumptions that he or she intends to make manifest or more manifest to an audience. Following the work of Grice (1957), most scholars maintain that interpersonal communication consists of the sender intending to cause the receiver to think or do something just by getting the receiver to recognize that the sender is trying to cause that thought or action. Thus, "communication is a complex kind of intention that is achieved or satisfied just by being recognized" (Levinson, 1983, p.16).

For instance, in Sam's narrative, we assume that his comment that *It (infertility) became a black hole for both of us* is preceded by a private mental act from which an intention to perform some behavior arises, namely to describe Sam's shared emotional experience with his wife in terms of being in a black hole. In making this metaphorical statement, Sam likely wishes for listeners (or, in this case, readers) to draw specific inferences about the meanings of his black hole statements, including, perhaps, inferences about his emotional state of mind (e.g., that he and his wife felt despair, outrage, loss of control, and so forth).

One question we address in this chapter is whether the communication of emotions is, at least sometimes, recognized as intentional on the part of speakers in everyday conversation. There are two parts to this issue that warrant close examination. First, can emotional communication be viewed as intentional in the sense of a person consciously and deliberately forming an intention to, say, open a window, and then getting up to actually fulfill the intention? Our argument is that emotions themselves are not the product of intentional thought processes. Instead, the idea of intentionally communicating emotions implies that people, or speakers, intentionally express their emotions through some set of displays that others are to recognize as part of speakers' overall communicative messages. These displays include facial expressions, body posture, tone of voice, and particular words. Although some emotional displays arise naturally as byproducts of people's autonomic nervous systems (e.g., flushed faces), and thus some information about emotional states "leak" out (Ekman, 1992), many others (e.g., posture, tone of voice, choice of words) are to an important degree under the control of the actor.

A second issue with intentions and emotions concerns whether emotions are the type of psychological behavior that can be communicated in the way that an idea can through language. Practically all of the cognitive science work on intentional communication focuses on how speakers verbally formulate, and listeners explicitly recover, communicative intentions that can be stated as well-formed propositions (Clark, 1996). Yet in many cases of interpersonal communication, never mind when reading literature, what is important is not the explicit propositions stated, but the more indeterminate nonpropositional meaning and affect that is expressed and understood (Gibbs, in press; Gibbs & Gerrig, 1989). Metaphor, for example, is a special communicative tool because it can, in some cases, create a sense of intimacy between speaker and listener that literal language is less able to do (Cohen, 1979; Gibbs & Gerrig, 1989). The intimacy associated with metaphor (including slang, idioms, proverbs) is partly centered around the sharing of emotions through aesthetic language, whose meaning is nearly impossible to describe in propositional form.

Our approach to emotional communication as a kind of nonpropositional meaning does not demand that the emotional message expressed must necessarily be explicitly enumerable and conscious in the mind of the speaker/actor. As Sperber and Wilson (1986) argued in the context of their *relevance theory*, a fundamental mistake in the field of pragmatics (i.e., the study of utterance interpretation) is to suppose that pragmatics "should be concerned purely with the recovery of an enumerable set of assumptions, some explicitly expressed, others implicitly conveyed, but all individually intended by the speaker" (p. 201). But

> [T]here is a continuum of cases, from implicatures which the hearer was specifically intended to recover to implicatures which were merely intended to make manifest, and to further modification of the mutual cognitive environment of speaker and hearer that the speaker only intended in the sense that she intended her utterance to be relevant, and hence to have rich, and not entirely foreseeable cognitive effects. (p. 201)

Consider again Sam's statement about the infertility becoming a black hole for him and his wife. Under the view of relevance theory, Sam did not necessarily have only one communicative intention in mind when he said what he did. He may have wanted listeners to specifically understand something about his experience by making reference to what it might conceivably be like to live, even temporarily, in a black hole. But Sam may also have aimed to make manifest in a wider sense other meanings that he didn't explicitly have in mind, as the terrible emotional consequences of living in black holes. His use of the black hole metaphor conveys some of the relevant emotions to his experience of infertility, such as despair, hopelessness, feeling out of control, anger, and so forth. Thus, Sam might have in mind a "cognitive content" without explicitly having in mind everything that might be entailed by that cognitive content in question. In many instances, speakers, like Sam, will agree, when asked, whether some interpretation of his or her utterance was appropriate, even if this reading might not have been exactly what he or she had in mind when the utterance was first spoken. Thus, if we questioned Sam as to his intentions in saying what he did, Sam could elaborate on his utterance by mentioning several possible meanings that he did not have firmly in mind at the time when he first spoke.

The main point here is that there is no reason not to expand the notion of an intended meaning in exactly this way so that speakers would, in fact, frequently accept as correct interpretations of their utterances various implicatures that they did not specifically have in mind when they originally framed the utterances in question. Among these are possible nonpropositional affective and emotional intentions that speakers may, to varying degrees, wish to make manifest by having listeners draw, again to varying degrees, different inferences about the speaker's emotional state of mind.

FIGURATIVE RHETORIC IN EVOKING EMOTIONS

Figurative language is used to express and evoke emotions in many kinds of discourse situations. Politics is a great place to see how figurative language may be employed specifically to evoke particular emotions, which may in turn influence a person's beliefs about some topic. Politicians are famous, or infamous, for their use of metaphorical language to evoke emotions. Consider the debate that took place in the U.S. Senate in January 1991 over whether the United States should take military action against Iraq for its invasion of Kuwait. An analysis of the *Congressional Record* of the 3-day debate in the Senate showed that metaphor was widely used by both Republican and Democrats to bolster their positions (Voss, Kennet, Wiley, & Engstler-Schooler, 1992). For instance, one Republican senator went to great lengths to evoke people's emotional response to Hussein by describing him in vivid metaphorical terms (Voss et al., 1992):

> Saddam Hussein is like a glutton—a geopolitical glutton. He is sitting down at a big banquet table, overflowing with goodies. And let me tell you—like every glutton, he is going to have them all. Kuwait is just the appetizer. He is gobbling it up—but it is not going to satisfy him. After a noisy belch or two, he is going to reach across the table for the next morsel. What is it going to

be? Saudi Arabia? He is going to keep grabbing and gobbling. It is time to let this grisly glutton know the free lunch is over. It is time for him to pay the bill. (p. 209)

Although it is not clear how effective these metaphors were in persuading other senators, or even whether these metaphors actually evoked specific images, associations, or emotion, it is clear that metaphors were used as reasons to support claims and to emphasize, concretize, or personalize particular issues (Voss et al.,1992). Experimental work has shown that metaphor can indeed significantly change people's attitudes toward various political and social topics (Bosman, 1987; Read, Cesa, Jones, & Collins, 1990). A significant part of metaphor's ability to persuade lies in its special ability to evoke particular emotions in listeners.

Metaphor is not the only rhetorical trope used to express and evoke emotions. Consider the following radio advertisement, sponsored by the California Department of Health Services, played in May 1998 on California radio stations. The radio spot is spoken in the voice of a 60-year-old man in a very sincere tone:

We, the Tobacco Industry, would like to take this opportunity to thank you, the young people of America, who continue to smoke our cigarettes despite Surgeon General warnings that smoking causes lung cancer, emphysema, and heart disease. Your ignorance is astounding, and should be applauded. Our tobacco products kill 420,000 of your parents and grandparents every year. And yet, you've stuck by us. That kind of blind allegiance is hard to find. In fact, 3,000 of you start smoking everyday because we tobacco folks tell you it's cool. [Starts to get carried away.] Remember, you're rebels! Individuals! And besides, you impressionable little kids are makin' us tobacco guys rich!! Heck, we're billionaires!! [Clears throat/Composes himself.] In conclusion, we the tobacco conglomerates of America, owe a debt of gratitude to all teens for their continued support of our tobacco products despite the unfortunate disease and death they cause. Thank you for your understanding. Thank you for smoking. Yours truly, The Tobacco Industry.

We're sure that listeners have different emotional reactions to this radio message. As three non-smoking adults, who decry the way the tobacco industry seduces children to smoke, we see great humor in the speaker's irony (e.g., *That kind of blind allegiance is hard to find*). Undoubtedly, smokers, and perhaps most teenagers, may not feel quite the same about this message as we did. But no matter how one responds emotionally, the irony here seems more effective in getting listeners' attention, and making them feel *something*, than would be the case if the speaker pleaded his case using literal speech.

METAPHOR IN SPEAKING OF EMOTIONS

There has been a great deal of research on the role of metaphor in people's talk of their own emotional experiences. Cognitive linguistic work has shown that talk of emotion is replete with metaphor (and metonymy)(Kovecses, 1986, 1990, this volume; Lakoff, 1987; Lakoff & Johnson,1980; Yu, 1998). A central claim of these studies is that human emotions are to a great extent conceptualized and expressed via metaphor grounded in embodied experience. Consider some of the ways people speak of anger (Yu, 1998, p. 51):

Those are inflammatory remarks.
She was doing a slow burn.
He was breathing fire.
Your insincere apology just added fuel to the fire.
After the argument, Dave was smoldering for days.
Boy, am I burned up.
Smoke was pouring out of his ears.

These expressions are not "dead" metaphors, but reflect the pervasive conceptual metaphor in which the abstract concept of anger is conceptualized in terms of heat, both within the human body and outside it. A more specific version of this general metaphor is the conventional metaphor that ANGER IS HEATED FLUID IN A CONTAINER, which gives to verbal expressions such as the following (Yu, 1998, p. 51):

You make my blood boil.
Simmer down.
I had reached the boiling point.
Let me stew.
She was seething with rage.
She got all steamed up.
Billy's just blowing off steam.
He flipped his lid.
He blew his top.

Our understanding of anger (the target domain) as heated fluid in a container (the source domain) gives rise to a number of interesting entailments. For example, when the intensity of anger increases, the fluid rises (e.g., *His pent-up anger welled up inside of him*). We also know that intense heat produces steam and creates pressure on the container (e.g., *Bill is getting hot under the collar* and *Jim's just blowing off steam*). Intense anger produces pressure on the container (e.g., *He was bursting with anger*). Finally, when the pressure of the container becomes too high, the container explodes (e.g., *She blew up at me*). Each of these metaphorical entailments is a direct result of the conceptual mapping of anger onto heated fluid in a container. Cognitive linguistic work on metaphors in emotion talk suggests, then, that there is a tight link between the specific ways people metaphorically conceptualize of their emotions and the language they use to express emotions. Thus, there is a significant difference between saying *Bill is getting hot under the collar* and *Bill is bursting with anger* that reflects something about the emotional intensity Bill might be experiencing.

Linguistic analyses of metaphor are supported by experimental studies looking at how people talk of their autobiographical experiences and emotions. Experimental studies have set up situations to investigate exactly when people use metaphor to speak of their emotions. Participants in one study were asked to give verbal descriptions for specific emotional states they had experienced and of activities in which they had engaged as a result of these remembered emotional states (e.g., happiness, pride, gratitude, sadness, fear, and shame) (Fainsilber & Ortony, 1987). According to the idea that metaphor allows one to express ideas that are difficult to talk about using literal language (the inexpressibility hypothesis; Ortony, 1975), people should be more likely to use metaphor and metaphorical comparisons to describe their subjective experiences of emotion than to describe the actions they took in response to the emotional experience (i.e., pounding a fist on a table to express anger). Moreover, according to the idea that metaphor allows one to speak in a concrete, vivid manner (the vividness hypothesis; Ortony, 1975), metaphorical language should be more prevalent in descriptions of intense as compared to mild emotional states.

Fainsilber and Ortony's (1987) study showed that people's descriptions of their emotional states contained more metaphorical language than did descriptions of their behavior. For instance, people described their negative emotional states with remarks such as *It was like someone had just dropped a bomb on me* and positive emotions with statements such as *It was like a very bright light was just shining outward.* Metaphor seemed especially useful to participants in expressing what was normally difficult to talk about using literal language. Metaphor was more frequently used to describe intense emotions than to talk of mild emotions. However, people did not use metaphor any more often for describing intense emotion than they did for milder emotion when simply talking about activities that resulted from experiencing a particular emotional state. These findings support the vividness hypothesis in that metaphor seems to be particularly suited for describing intense emotions. Additional experimental evidence demonstrated that creative writers employ metaphor quite frequently when asked to describe their own feelings and actions (Williams-Whitney, Mio, & Whitney, 1992).Most generally, metaphor is not simply used as linguistic ornament, but serves an indispensable communicative function.

More recently, Fussell (1992) asked undergraduates to write descriptions of specific instances in which they had experienced mild and intense feelings of anger, sadness, happiness, and pride. Participants produced figurative language for affective, cognitive, and bodily responses in describing their autobiographical experiences of anger. Affective responses included *I felt like a coiled spring* and *I don't want to blow my top*, cognitive responses included *I want to put somebody through the wall into the next room* and *I had a desire to crush the others*, and bodily responses included *My stomach was twisted in knots* and *I feel like I'm going to burst*. The data also showed that participants used figurative language significantly more often when describing intense, as opposed to mild emotional experiences, especially for sadness and happiness. Overall, most of the figurative expressions people produced were conventional or frozen, but there were still a significant number of novel expressions (e.g., *My mind was seething and boiling* and *My entire insides seemed ready to burst*).

Fussell and Moss (1998) went on to show that people frequently use metaphor when describing other people's emotions. Participants watched several movie scenes in which the characters were sad or depressed. Afterwards, participants described the characters' emotions to someone who had not seen the movie clips. Most notably, the speakers captured the nuances of the depicted states of depression using figurative, but not literal, language. Thus, literal phrases, such as *sad, angry*, or *depressed* were used evenly when talking about all the different movie scenes. But idioms and metaphors were carefully tailored to specific scenes to capture the specific emotions of the movie characters. For instance, the character in the scene from *Winter People* was described as *dazed, shocked*, and *in a trance*, and the character in *Steel Magnolias* was described in terms like *breaking down, going ballistic*, and *going crazy*. The data support the idea that figurative language is special, in part, because it differentiates between complex variations of a single emotional state in a way that literal messages cannot.

One of the interesting findings from Fussell and Moss (1998) was that speakers did not use figurative expressions in lieu of literal ones, but in addition to them. This result is consistent with what is seen in other narratives. For example, in Sam's narrative about his wife, he mentioned that *she was constantly depressed*, but again elaborates on this literal description by talking about the wife being in *a tremendous black hole* and that *she couldn't get out of it*. In both Sam's narrative and the narratives from Fussell and Moss speakers clearly recognize that the literal emotion term by itself is inadequate to express the complexity and intensity of either their own or someone else's emotional state(s). Thus, choosing to speak figuratively to express or describe emotions doesn't merely replace literal speech, but complements or elaborates on it.

Psycholinguistic studies have examined in more detail why it is difficult to literally express one's emotions. Consider a situation where a speaker is angry at a listener. Why might the speaker say *You make me blow my stack* opposed to either the idiom phrase *You make me want to tear your head off*, or even the literal expression *You make me very angry*? Research has shown that people's knowledge of the metaphorical links between different source and target domains provides the basis for the appropriate use and interpretation of conventional, verbal metaphors in particular discourse situations (Nayak & Gibbs, 1990). Participants in one study, for example, gave higher appropriateness ratings to *blew her stack* in a story that described the woman's anger as being like heat in a pressurized container whereas *bit his head off* was seen as more appropriate in a story that described the woman's anger in terms of a ferocious animal. *Bite your head off* makes sense because people can link the lexical items in this phrase to the conceptual metaphor ANGRY BEHAVIOR IS ANIMAL BEHAVIOR. An animal jumping down a victim's throat is similar to someone shouting angrily.

On the other hand, people understand the figurative meaning of *blow your stack* through the conceptual metaphor ANGER IS HEATED FLUID IN A CONTAINER where a person shouting angrily has the same explosive effect as does the top of a container blowing open under pressure. Thus, readers' judgments about the appropriateness of an idiom in context were influenced by the coherence between the metaphorical information depicted in a text and the conceptual metaphor underlying an idiom's figurative meaning. Even though we may have many idiomatic phrases that refer to a single concept (e.g., anger), some of these phrases may be motivated by different underlying conceptual metaphors (e.g., *Blow your stack* vs. *bite your head off*). Because our emotions are often understood via multiple, and sometimes contradictory, metaphors, it is not surprising that we have so many different kinds of conventional phrases to reflect sometimes subtly different aspects of our everyday emotional experience.

Various experimental work shows that an emotion such as anger is prototypically understood to be composed of a sequence of events beginning with a set of "antecedent conditions," a set of "behavioral responses," and a set of "self-control procedures" (Shaver, Swartz, Kirson, & O'Connor, 1987). The temporal organization of many emotions like anger is reflected in the language used to refer to these emotions (Lakoff, 1987). Thus, English has different idioms fitting each of the three stages for our concept of anger. The antecedent conditions (Stage 1) for anger concern a sudden loss of power, status, or respect, ideas exemplified by idioms such as *eat humble pie*, *kick in the teeth*, and *swallow one's pride*. The behavioral conditions (Stage 2) for anger concern people's behavioral responses to the emotion, an idea best reflected in idioms such as *getting red in the face*, *getting hot under the collar*, or *blowing your stack*. The self-control procedures (Stage 3) concern an individual's efforts to maintain composure, an idea that is best reflected in idiomatic phrases such as *keep your cool*, or *hold your temper*.

Idiomatic phrases that refer to the same temporal stage of a conceptual prototype appear to be highly similar in meaning. For example, *do a slow burn* and *blow your stack* both mean "to get angry". It is likely that similarities in the figurative meanings of idioms arise from their referring to the same temporal stage of a conceptual prototype (e.g., Stage 2). This possibility implies, contrary to the view of traditional theories, that idioms are linked together in the mental lexicon on the basis of the temporal stage of the concept to which each idiom refers. Thus, idioms referring to the same temporal stage (e.g., *blow your stack* and *do a slow burn*) may be more closely linked (with all the phrases sharing the same semantic space or field) than are idioms that express different temporal aspects (e.g., *do a slow burn* and *hold your temper*).

People's sensitivity to similarity in the meanings of idioms may be based, in part, on whether these phrases express the same temporal aspect of the prototype to which they refer (e.g., *blow your top* and *flip your lid*). One study showed, in fact, that people's judgments of the similarity in the figurative meanings of two idioms were influenced by the temporal properties of emotion prototypes (Nayak & Gibbs, 1990). Idioms referring to the same stage of an emotion (e.g., *play with fire* and *go out on a limb*) were judged to be more similar in meaning than idioms (e.g., *play with fire* and *shake in your shoes*) referring to different temporal stages. These data demonstrate that idioms referring to emotions are not only categorizable on the basis of the temporal stage within a prototype, but these categories are also systematically related to one another in a strict temporal sequence. Idioms with similar meanings may be clustered around the emotion prototype to which they refer, with individual idioms being temporally linked much like our temporal knowledge of the events of mundane concepts such as "going to a restaurant" (cf. Schank & Abelson, 1977).

One important consequence of the idea that figurative language like idioms reflects the metaphorical mappings between source and target domains is that idioms have more complex meanings than do their typical literal paraphrases. These idiomatic meanings can be partly predicted based on the independent assessment of people's folk understanding of particular source domains that are part of the metaphorical mappings that motivate these idioms' interpretations. That is, by looking at the inferences that arise from the mapping of heated fluid in a container onto the experience of anger, one can make specific predictions about what various idioms motivated by ANGER IS HEATED FLUID IN A CONTAINER actually mean. More importantly, though, is that by understanding the motivation for why idioms mean what they do, we can see exactly why idioms, as one type of figurative language, are well suited to talk about specific emotional experiences.

The results from several experiments explicitly showed that people's understanding of idiomatic meaning reflects the particular entailments of their underlying conceptual metaphors (Gibbs, 1992). Participants in a first study were questioned about their understanding of events corresponding to particular source domains in various conceptual metaphors (e.g., the source domain of heated fluid in a container for ANGER IS HEATED FLUID IN ACONTAINER).For example, when presented with the scenario of a sealed container filled with fluid, the participants were asked about causation (e.g., *What would cause the container to explode?*), intentionality (e.g., *Does the container explode on purpose or does it explode through no volition of its own?*), and manner (e.g., *Does the explosion of the container occur in a gentle or violent manner?*).

Overall, the participants in this study were remarkably consistent in their responses to the various questions. To give one example, people responded that the cause of a sealed container exploding its contents out is the internal pressure caused by the increase in the heat of the fluid inside the container, that this explosion is unintentional because containers and fluid have no intentional agency, and that the explosion occurs in a violent manner. More interesting, though, is that people's intuitions about various source domains maps onto their conceptualizations of different target domains in very predictable ways. Thus, other studies in this series showed that when people understand anger idioms, such as *blow your stack*, *flip your lid*, or *hit the ceiling*, they inferred that the causes of the anger is internal pressure, that the expression of anger is unintentional, and is done is an abrupt violent manner. However, people do not draw inferences about causation, intentionality, and manner when comprehending literal paraphrases of idioms, such as *get very angry*. Literal phrases, such as *get very angry*, are not motivated by the same set of conceptual metaphors as are specific idioms such as *blow your stack*. For this reason, people do not view the meanings *blow your stack* and *get very angry* as equivalent despite their apparent similarity. *Blow your stack* has a complex figurative meaning (e.g., getting angry when there is intense pressure, when the anger is expressed unintentionally, and done so in a violent manner) that captures subtle nuances of people's anger experiences. Literal phrases, such as *get very angry* are vaguer and don't capture these subtle aspects of emotional experience. Thus, an important reason why figurative language is special in emotional communication is because it tighter reflects something about people's ordinary conceptualizations of their complex emotional experiences.

EMOTIONAL REACTIONS TO METAPHOR

How do listeners respond emotionally to figurative as opposed to literal phrases? Turner and Gibbs (in preparation) recently conducted a study to investigate whether speakers convey more emotional intensity and induce stronger emotional reactions in listeners when they describe their emotions using metaphorical language. Our hypothesis was that metaphorical language would convey more emotional intensity and evoke more emotional reactions in listeners than would literal speech precisely because of the special nature of metaphorical language to communicate a wide range of meanings. We also investigated whether novel metaphors are more effective than conventional metaphors at conveying emotional intensity. According to Lakoff and Turner (1989), metaphors are conventional when they are grounded in our everyday experience and we use them automatically and effortlessly. For example, saying *I was totally shaken* is a conventional way to communicate fear whereas *I was a bowl of quivering jello* is a novel way to say that one was frightened.

Why might novel metaphor communicate more emotional intensity than conventional metaphor? One argument for the superior ability of metaphor, in general, as a means of conveying emotion is that metaphor arises from embodied experience (see Johnson, 1987; Gibbs, 1994). The expression *I totally exploded* involves the entire body and implies action, whereas saying *I got really angry* describes emotional experience in a more abstract, less embodied way. Understanding novel metaphors also, in many cases, requires additional cognitive effort to understand than is required of either conventional metaphors or literal speech (Gibbs, 1994). The extra effort not only goes toward understanding what the speaker generally meant, but also to comprehend something about the nuances of that meaning, as well as the aesthetic nature of the speaker's emotional experience. To say *I was a bowl of quivering jello* when fearful provides a precise embodied description of the speaker's emotional state and presents a vivid, almost poetic, depiction of what it was like for the speaker to feel this fear. Thus, the creation of a novel metaphor should make this phrase especially useful for expressing a speaker's emotional intensity and to evoke emotion on the part of the listener (e.g., vicarious fear, sympathy for the speaker, etc).

There is some preliminary evidence to suggest that there is a relationship between novel metaphor use and emotional intensity. Fainsilber and Ortony (1987) found that conventional metaphorical expressions were used eight times as often as novel metaphors to describe emotional states but the ratio of novel to conventional metaphor use was greater when their participants described intense emotions (12%) than when they depicted mild emotions (8%). Similarly, Whitney-Williams, Mio, and Whitney (1992) found that writers used more novel metaphors when describing their own intense feelings. Only participants that were categorized as experienced writers used more novel metaphors when describing the feelings of other people. One explanation for this difference is that people want to appear unique when describing themselves, but use more stereotypical and less cognitively taxing depictions when describing others (see Whitney-Williams et al., 1992). This explanation is similar to Gerrig and Gibbs' (1988) contention that speakers use creative utterances to individuate personal experiences from general experiences in the lexicon. Novel metaphorical expressions may also establish greater intimacy between speaker and listener because they rely on, and help establish, highly specific common ground between the participants.

The research discussed here involved having participants generate descriptions of emotional states that were then coded for the number and type of metaphors used. Turner and Gibbs (in preparation), on the other hand, looked at the effect metaphorical language has on the listener. Twenty-three undergraduate students listened to an audiotape containing 24 scenarios. Each tape contained six scenarios for each of four emotions (anger, fear, happiness, and sadness). Of the six scenarios, two ended with literal statements, two with conventional metaphorical expressions, and two with novel metaphorical expressions. The tapes were counterbalanced so that, over the six versions of the tape, each scenario was paired with each of the endings for that particular emotion. Within each tape, the order of scenarios was randomized.

Each scenario and ending were approximately the same length. The endings for fear and anger were stated using past tense because they described a specific instance of emotion. Because happiness and sadness are less transient mood states, the endings for these scenarios were expressed in present tense. The gender of the characters varied across the scenarios. An example of one scenario (for anger) is the following:

> My friend's roommate borrowed her car without asking her. She has a cobalt blue 1997 Jetta in perfect condition. He was in a hurry and parked the car in a tight space in a store parking lot and when he returned, the car had a huge dent in the fender. When he got home with the car, my friend noticed the dent. She told me ...
>
> *I was really angry.* (literal statement)
> *I hit the ceiling.* (conventional metaphor)
> *I was a live grenade.* (novel metaphor)

The tapes were recorded by a 22 year-old female speaker, who used a monotone voice to avoid conveying any emotional intensity other than that communicated by the words themselves. Participants were given a sheet containing 9-point Likert-type scales on which they were asked to rate the emotional intensity the speaker felt given what he or she said and the intensity they felt given what the speaker said. They were instructed to fill out the scales during a pause in the tapes that occurred after each scenario.

Analyses of the participants' ratings of speakers' emotions revealed significant differences among the four emotions and the three linguistic statements. Specific comparison tests showed that happiness (6.74) was rated as significantly more intense than the other emotions. Anger (5.95) and fear (5.92) were rated as significantly more intense than sadness (5.34). More importantly, specific comparison tests showed that both novel (6.24) and conventional (6.12) metaphorical expressions were rated as reflecting more intense emotion than were literal statements (5.58). The two types of metaphorical statements did not significantly differ from each other.

A separate analysis on participants' feelings of intensity demonstrated significant differences among the four emotions and the three linguistic statements. Specific comparison tests revealed that happiness (5.88) was rated significantly more intense than sadness (5.19). Anger (5.42) and fear (5.59) did not significantly differ from happiness or sadness. Analysis of the statements showed that conventional metaphor (5.90) and novel metaphors (5.47) induced more emotional intensity than literal statements (5.19), but the difference was not statistically reliable.

This overall pattern of findings is consistent with the idea that there is a relationship between metaphorical language and emotional intensity. The results of this study extend previous findings that people are more likely to use metaphor to express emotional states, by determining that listeners will also infer greater emotional intensity on the part of a speaker who uses metaphorical language. Furthermore, people appear to experience more emotional intensity when they hear metaphorical language. It is interesting that participants attributed more emotional intensity to a speaker who used novel, rather than conventional, metaphorical expressions (but this difference, again, was not statistically significant). Our findings might be explained, in part, by the results of the Williams-Whitney, Mio, and Whitney (1992) study in which both experienced and novice writers used more novel metaphors than conventional ones when describing the emotional experience of another person. Experienced writers may have a greater ability to role play and be empathic with characters. Given that our empirical findings required participants to be observers, it may have demanded a greater amount of empathy than if the participants were personally involved in the scenarios.

We found that participants consistently rated happiness as having greater intensity than the three negative emotions. It may be that the scenarios we wrote for happiness were more engaging, or that participants were more attracted to positive scenarios or that they were not willing to empathize or be put in a negative mood by the experimental material. Yet Fainsilber and Ortony (1987) also found that metaphors were particularly likely to be used to describe intense positive emotions.

The results generally supported the hypotheses that metaphorical language both reflects and induces greater emotional intensity than literal language. The results suggest, but do not unequivocally support, the hypothesis that novel metaphors convey more emotional intensity on the part of the speaker. The hypothesis that novel metaphors induce more emotional intensity in listeners than conventional metaphors was not confirmed.

IRONY IN SPEAKING OF EMOTION

Like metaphor, irony is often used to express emotion. Consider these exchanges between the husband, George, and his wife, Martha, in Edward Albee's play *Who's Afraid of Virginia Woolf?* (Bollobas, 1981).

Martha:	Why don't you want to kiss me?
George:	Well dear, if I kissed you, I'd get all excited... I'd get beside myself, and I'd take you, by force, right here on the living room rug, and then our little guests would walk in, and ... well, just think what your father would say about that
	* * * *
Martha:	It's the most ... life you've shown in a long time.
George:	You bring out the best in me, baby.
	* * * *
Martha:	(...) You have a poetic nature, George ... a Dylan Thomas-y quality that gets me right where I live.

These statements mock the addressee by expressing the opposite of what is meant (e.g., "Martha brings out the worst in George," "Kissing Martha would not get George excited," and "George does not have a poetic nature.").

Consider the following conversation between two parents (Bill and Pat) and their teenage son (Grant)from the documentary film series *The American Family* (Gibbs, 1994).The conversation takes place in the family's backyard by their swimming pool.

(1)	Bill:	Come over here a little closer...I think
(2)	Grant:	Well, I'd rather stay out of this
(3)	Bill:	You...want to stay out of swinging distance
(4)	Grant:	Yeah, I don't want to hurt you
(1)	Bill:	Well...I mean...I was talking with your mother, ya know, and I told her that you weren't interested in doing any more work, ya see...and I don't blame you... I think that's a very honest reaction, there's nothing wrong feeling, umm...its a natural thing to do not wanting to work
(2)	Grant:	No, ah...it's not that I don't want to work, it's just ah..
(3)	Pat:	What kind of work did you have in mind, Grant? Watching the television, and listening to records
(4)	Grant:	I don't need your help mom
(5)	Pat:	Playing the guitar, driving the car
(6)	Grant:	Ah..
(7)	Pat:	Eating, sleeping
(8)	Bill:	No, ah, listen Grant, you are a very good boy, we're very proud of you
(9)	Grant:	Yeah, I know you are
(10)	Bill:	No, we are...you don't give us any trouble ya know
(11)	Grant:	Well, you sure are giving me a hell of a lot

(12) Bill:	Well that's my job I think...if, ah, I don't why nobody else will and that's what I'm here for you...is to kind of see that you get off to a good start. Lucky you may be with the deal, that's my job is to see that you get to see how life's going to be
(12) Grant:	Yeah
(13) Bill:	and, ah, if I don't then nobody else will...a lot of kids go around don't ever have that privilege of having a mean old man
(14) Grant:	Yeah, sure is a privilege too (pp. 375-376)

This conversation is not atypical of how many American families talk, particularly when parents converse with their teenage children. These participants employ figurative language, especially irony and sarcasm, to good effect not only to indirectly convey various communicative intentions, but also to assert their own figurative understanding of the topics under discussion. For example, Grant says in (4) that he doesn't want to hurt his father and means this literally, but intends in a jocular way to convey the pretense that he *could* hurt his father should they get into a physical fight. This utterance fulfills the need for father and son to defuse what must be mutually recognized as a potentially uncomfortable situation of the parents criticizing Grant for his unwillingness to work. Later, in (7), (9), and (11), Pat sarcastically echoes Grant's putative belief that watching television, listening to records, playing the guitar, and so on, constitute meaningful work activities. Pat's sarcasm reflects her ironic understanding of Grant's claim that he wants to work while at the same time doing little other than watch TV, listen to records, play the guitar, and so on. Finally, in (14) Grant sarcastically comments on the privilege he feels in having *a mean old man* for a father.

Both conversations from *Who's Afraid of Virginia Woolf?* and *An American Family* illustrate how people's conceptualizations of situations as ironic leads to their speaking ironically or sarcastically (Gibbs, 1994). These examples highlight the idea that when a person's communication is sufficiently discrepant from what is expected, or hoped for, the violation will be seen as arousing or disturbing, directing some attention away from the ostensible topic at hand and toward the violation and violator (Burgoon, 1993). In this manner, ironic speech not only reflects people's ironic conceptualizations of experience, but should also be especially useful in provoking emotions in others.

EMOTIONAL REACTIONS TO IRONY

How does irony influence people's emotional reactions to what speakers say? Many scholars have argued that irony is special because it represents a mode of intellectual attachment and that "irony engages the intellect rather than the emotions" (Walker, 1990, p. 24). The modernist tradition, best represented by writers like I. A. Richards and T. S. Eliot, equates irony less with provoking emotions than with equilibrating them. But irony's ability to mock, attack, and ridicule, provoking embarrassment, humiliation, even anger, suggests the opposite of this conclusion. Although irony has a decidedly intellectual edge, it also implies a judgmental attitude with a strong affective, emotional dimension. Irony may, somewhat paradoxically, signal detachment, or "the cutting edge of not caring" (Austin-Smith, 1990, p. 51), yet also can reflect deep emotion and evoke a range of affective responses.

Psycholinguistic research paints a mixed picture of the relationship between irony and affective responses. Dews and Winner (1995) proposed the "tinge hypothesis" that ironic criticism automatically reduces the amount of condemnation that listeners experience. By stating literally positive words in an ironic criticism (e.g., A *fine friend you are!*), the speaker insures that listeners will interpret what is said in a more positive manner. Indeed, the results of several studies show that critical ironic statements are rated as less severe than literal ones (Dews & Winner, 1995). On the other hand, Colston (1997) demonstrated that ironic criticism in many cases actually enhances, rather than dilutes, condemnation and criticism. Neither of the mentioned studies, however, specifically examines people's emotional reactions to different forms of irony in comparison to literal speech.

How might different forms of irony influence listeners' emotional reactions to what is communicated? Imagine yourself in the following situation: One day while parking at work your car splashes mud on Mary. Mary walks over to your car while you are getting out. You look at Mary and ask why her clothes are such a mess. Mary looks back at the mud puddle in the road and answers: *You splashed mud on me with your car.*

Mary's utterance might very well make you feel guilty for doing what you, perhaps unintentionally, did to dirty her clothes. However, Mary's comment, although quite factual, doesn't convey much about her own attitude or emotion in regard to your act. She might simply be calling attention to what you did while forgiving your actions, or she might actually be displeased or quite angry with you.

Suppose, instead, that Mary actually uttered the sarcastic remark *Thanks a lot for giving me a bath.* Once again, you might feel guilty on being alerted to your misdeed. Yet by virtue of informing you about what you did in a sarcastic manner, you are likely to recognize that Mary appears rather angry. Sarcasm is considered especially appropriate for conveying a person's hostile attitude toward, or ridicule of, some other individual, usually the addressee (Gibbs, 1994; Lee & Katz, 1998). Consequently, being the object of another person's sarcasm might provoke intense emotional reactions.

Mary might also have uttered the rhetorical question *Do you ever know where you're going*? Rhetorical questions not only indirectly convey Mary's negative attitude toward you for splashing her with mud, but they directly question your general ability to move about in the world without causing harm to others. Hearing a rhetorical question like *Do you ever know where you're going*? reveals the speaker's anger toward you and feels like a direct personal attack, perhaps making you feel guilty or even angry, especially if what you did was unintentional and a rare occurrence.

Mary might comment about your splashing mud on her in a variety of other ironic ways, each of which may reflect different emotional states of mind. If she uttered an understated reproach such as *You might want to drive a bit slower*, Mary clearly states her desire for you to be more careful, but also expresses this from a less hostile point of view, perhaps reflecting her understanding that your act was unintentional. Being the target of understated criticism like this might make you feel guilty, but probably far less so than if she said something sarcastic or if she uttered an angry rhetorical question. In contrast, if Mary produced a hyperbolic comment like *You soaked me to the bone*, you might get the strong sense by her exaggeration that Mary was quite upset, although, again, not quite as angry as if she'd made a sarcastic remark. Finally, Mary might only have stated *I just put on these clothes*, a comment that makes it more difficult to predict exactly what she was feeling. This statement is indirect, it makes no mention of the problem or offense, but is likely to reflect a negative reaction or feelings.

These examples illustrate that irony involves a wide range of rhetorical types, each of which might function to evoke slightly different emotional reactions in listeners. Moreover, speaking ironically, in one of the ways mentioned, may signal quite a different conceptualization of some event than if someone made literal comments, such as if Mary, in this scenario, said *You have splashed mud on me* or *What you did makes me angry*. There is a growing body of psycholinguistic evidence that people achieve a complex set of social and communicative goals by speaking ironically (Colston, 1997; Dews, Kaplan, & Winner, 1995; Kreuz, Long, & Church, 1991; Kreuz & Roberts, 1995; Kumon-Nakamura, Glucksberg, & Brown, 1995), including being humorous, acting aggressively, achieving emotional control, elevating one's social status, expressing attitudes, provoking reactions, mocking others, and muting the force of one's meaning. Emotions appear to underlie most or all of these goals. Most theorists are aware that emotions are important with verbal irony, for example, discussion of surprise or violated expectations occurs with striking regularity (e.g., Dews et al., 1995; Haverkate, 1990; Kumon-Nakamura et al., 1995).

Leggitt and Gibbs (2000) specifically examine the particular patterns of emotions that are communicated and evoked by different ironic remarks in conversations. We thought that people might have quite different emotional reactions depending on the phrasing of ironic statements, even though similar information is conveyed. Moreover, we hypothesized that people would experience different emotional reactions to various ironic statements than to literal ones, and, should actually feel more intense emotions having heard ironic remarks than literal ones, contrary to the muting hypothesis. We investigated these questions with a set of stories that ended in ironic statements. Questionnaires were given to three groups of participants, and different instructions were given to each group. The first group was asked to rate how they would feel on hearing each statement, the second group rated how the speaker felt when making each statement, and the third group rated how the speaker wanted the addressee to feel on making each statement. We included a wide range of ironic statements, including hostile ones such as sarcasm (e.g., *You are a real professional*) and rhetorical questions (e.g., *Why don't you just eat it all?*), less confrontational ones such as satire (e.g., *You must expect me to wear an apron*) and understatement (e.g., *You must have been a bit hungry*), plus overstatement (e.g., *You have walked by a million times*), irony not directed at the addressee (e.g., *And I rushed to be on time*), and non-ironic statements. The more that a type of statement is associated with distinct emotional reactions, the more the support for the claim that verbal irony is chosen to affect emotions. The greater the similarity between the speaker's intentions and the addressee's reactions, the more the support for the claim that speakers affect emotions because of their intentions. If language is chosen for emotional effects there should be a logical relationship between the emotions of the speaker and the addressee. For example the speaker generally acts in control and has power, but the addressee is vulnerable and threatened and so should feel more defensive.

The results of the first study ("How would you feel?") indicate that people feel different emotions on hearing various types of ironic statements. Sarcasm, rhetorical questions, and overstatement all evoked similar and quite negative reactions whereas understatement and satire evoked relatively neutral reactions. Nonpersonal irony evoked a lower degree of negative and hostile emotions than the other statements. In the second study ("How does the speaker feel?"), the participants rated non-personal irony as quite positive and nonhostile, speakers of satire were seen to feel more positive emotions and a lower degree of hostile emotions, and speakers of understatements were seen to feel relatively neutral emotions. Speakers of sarcasm and rhetorical questions appeared to feel a wide range of negative emotions, but speakers of overstatements appeared to feel relatively neutral emotions. In the third study ("How does the speaker want you to feel?") the participants perceived more positive intentions and a lower degree of negative intent with non-personal irony, and a greater desire to evoke positive emotions with satire. Understatement again was perceived as conveying a relatively neutral intent. Sarcasm and rhetorical questions were perceived as having very negative intent. Overstatements were again perceived to be uttered with relatively neutral emotional intentions, which also contrasts sharply with the strongly negative reactions of the addressees.

Correlational analyses were used to compare the types of irony with the other types of expressions. In all three experiments, statements fell into two categories, and the categories were consistent with the exception of overstatement. The addressees felt about the same when hearing overstatement, sarcasm, and rhetorical questions, and these were opposed to nonpersonal irony, understatement, and satire. Participants rated speakers' feelings about the same when using sarcasm or rhetorical questions, but rated speakers' feelings differently when nonpersonal irony, overstatement, understatement, and satire were used. Judgments of speakers' intentions revealed that participants felt speakers were conveying similar feelings to addressees when using sarcasm and rhetorical questions, but different feelings when using irony, overstatement, understatement, and satire. In general, the reaction to a statement seems to most strongly depend on the degree that the speaker directly challenges an addressee or makes a big deal of an issue. Overstatement was an exception to the general pattern; participants clearly felt that the speaker intended to get a positive reaction from the addressee, yet the addressee reacted as if the statements were very hostile and sarcastic. The phrasing of overstatement (e.g., *You have walked by a million times*) seems responsible for the ambiguity; the speaker might initially appear to make a big deal out of the problem but may often be joking.

Factor analysis was used to detect broader trends by averaging across all statements and contexts. This procedure allowed eight categories of emotions to be simplified into three related groups. In our stories the speakers were rated as feeling negative-aggressive emotions (e.g., anger, disgust) most strongly, positive emotions (e.g., happiness, amusement) second-most strongly, and negative-defensive emotions (e.g., fear, anxiousness) were least important. Speakers were viewed as wanting addressees to feel negative-aggressive emotions most strongly, to feel negative-defensive emotions second, and finally, to feel positive emotions. The addressees felt negative-aggressive emotions most strongly, negative-defensive emotions second, and positive emotions were least important. This pattern was consistent with our expectation that the speaker and addressee would feel logically related emotional states: The speakers generally desired to fix a problem or right a wrong when making a statement. The addressees' feelings did not precisely match the speakers' intentions: both negative-defensive and positive emotions were intended to be more important than what the addressees actually felt, and negative-aggressive intentions were less important than the addressees' feelings. This pattern supports the general claim that speakers choose a statement to affect emotions, but the differences imply that people do not have tight control over their reactions.

The best way to understand the relationship between the statements and emotions is to compare each statement across the three studies. Sometimes speakers are seen to use emotions intentionally and sometimes emotions might better be seen as out of control. With nonpersonal irony, understatement, and rhetorical questions, the speaker was seen to feel a particular way, was seen to want the addressee to feel about the same, and the addressees' feelings were consistent with the speakers' feelings and intentions. For example, rhetorical questions involved high levels of negative emotions across the studies whereas nonpersonal irony always involved low levels of negative emotions. These patterns support the claim that speakers intentionally communicate emotions and that emotions have predictable effects. With satire the speaker was seen to feel positive emotions and to intend a positive reaction, but the addressees' reactions were less positive than intended. This pattern is also consistent with the claim that speakers communicate emotions intentionally because this project studied hostile and negative encounters where sarcasm and rhetorical questions might be appropriate. The reactions to satire appear to result from situations where it would be difficult to feel positive emotions at all; for example, a person who has just made a mistake will tend to feel anxious, embarrassed, or angry.

There were two notable exceptions to the general pattern of the speakers' feelings matching their intentions and the addressees' reactions. Sarcastic speakers were seen to strongly feel negative emotions and to a limited degree want the addressee to feel the same, but the addressees felt negative emotions to a greater degree than was intended. It may be that sarcastic speakers appeared to display their own bad feelings rather than appearing to affect the addressees' emotions. Sarcastic speakers might be having a temper tantrum or might be acting like they are having a temper tantrum. With overstatement the speakers were seen to feel fairly neutral, were seen to want the addressees to feel fairly neutral, but the addressees reacted very negatively. The exaggerated phrasing probably misled the addressees to feel they had done something wrong whereas the speakers might have appeared to be joking. In this case it seems that the emotions of the addressees "got out of control." Overall, the evidence indicates that ironic statements are generally chosen because of how they affect addressees' emotions, but the exceptions show that emotions involve a degree of roughness and uncertainty about intentions.

CONCLUSION

We have looked at what's special about figurative language in emotional communication by investigating why people use different tropes to convey and evoke emotions. We approached this question from two perspectives: (a) communicative intentions when discussing emotions, and (b) conceptualizations of emotion. Speakers' intentions were addressed with theoretical and empirical evidence. We've argued, adopting the perspective of relevance theory, that speakers may not be consciously aware of all the cognitive effects of an utterance at the time they utter it. By using figurative language, they are able to make manifest an array of meanings. In a mutual cognitive environment, a listener can hear an utterance that a speaker intended to be rich and relevant and can make new interpretations in line with the speaker's intentions.

We hypothesized that figurative language can convey subtle nuances of meaning in a way that literal language cannot. In particular, different figurative expressions closely reflect people's figurative, often metaphorical, conceptualization of emotional experience. A variety of linguistic and empirical students support this conclusion. Furthermore, one shade of meaning that figurative language may communicate is intensity of emotion. We presented one study that offers preliminary evidence that metaphorical language reflects and conveys greater emotional intensity. A second study, which looked at various types of ironic utterances, found that ironic statements are more emotion-provoking than corresponding literal statements. This study centered on two hypotheses: (a) the more a type of statement is associated with a distinct emotional reaction, the more support for the idea that verbal irony is used when speakers want to affect listener's emotions, and (b) the more similar the intentions of the speaker and the reactions of the addressee, the more likely it is that speakers' intentions have an effect on listeners' emotions. Results indicated that when ironic statements are used, speakers' feelings match their intentions and addressees' reactions, with the exception that the use of sarcasm and overstatement led listeners to have more negative reactions than the speaker intended. Across the two studies, it appears that figurative language is used in great part for emotional effect.

Emotion scholars face several distinct challenges in future research First, greater attention needs to be given to the on-line character of producing and understanding figurative language given particular emotions. When someone is experiencing a specific emotion (and not just remembering it), is that person really likely to use particular kinds of figurative expressions in talking about that emotion? Do listeners of figurative language actually experience specific emotions (and not just experience them when asked to consciously reflect and write about them)? Finding the answers to these questions may be difficult given the limitations of current experimental methods for assessing on-line experience for specific emotions. We hope to extend some of the studies reported in this chapter to examine on-line emotional reactions to different kinds of figurative language. Beyond this, we think that a much wider range of figurative language needs to be examined in future studies. As pointed out in Gibbs (1994), there are many kinds of figurative language (e.g., metonymy, proverbs, oxymora, euphemism, slang), each of which might have their own respective special pragmatic and emotional properties. Finally, scholars need to focus more on the social nature of emotional experience. Rather than asking people to read or listen to almost anonymous speakers or writers, we need to look at how people respond emotionally in more realistic social situations when they have intimate interpersonal relationships with their conversational participants.

REFERENCES

Austin-Smith, B. (1990). Into the heart of irony. *Canadian Dimension*, *27*, 51-52.

Becker, A. (1997). *Disrupted lives*. Berkeley: University of California Press.

Bollobas, E. (1981). Who's afraid of irony? An analysis of uncooperative behavior in Edward Albee's "Who's Afraid of Virginia Woolf?". *Journal of Pragmatics*, *5*, 323-334.

Bosman, J. (1987). Persuasive effects of political metaphors. *Metaphor and Symbolic Activity*, *2*, 97-113.

Burgoon, J. (1993). Interpersonal expectations, expectancy violations, and emotional communication. *Journal of Language and Social Psychology, 12,* 30-48.

Clark, H. H. (1996). *Using language*. New York: Cambridge University Press.

Cohen, T. (1979). Metaphor and the cultivation of intimacy. In S. Sacks (Ed.), *On metaphor*. (pp. 1-10). Chicago: University of Chicago Press.

Colston, H. (1997). "I've never seen anything like it": Overstatement, understatement, and irony. *Metaphor and Symbol,12*, 43-58.

Dews, S., Kaplan, J., & Winner, E. (1995). Why not say it directly? The social functions of irony. *Discourse Processes,19*, 347-367.

Dews, S., & Winner, E. (1995). Muting the meaning: A social function of irony. *Metaphor and Symbolic Activity*, *10*, 3-19.

Ekman, P. (1992). *Telling lies: Clues to deceit in the marketplace, politics, and marriage*. New York: Norton.

Fainsilber, L., & Ortony, A. (1987). Metaphorical uses of language in the expression of emotion. *Metaphor and Symbolic Activity*, *2*, 239-250.

Fussell, S. (1992). *The use of metaphor in written descriptions of emotional states*. Unpublished manuscript. Carnegie-Mellon University.

Fussell, S., & Moss, M. (1998). Figurative language in emotional communication. In S. Fussell & R. Kreuz (Eds.), *Social and cognitive aspects of interpersonal communication*.(pp. 113-143). Mahwah, NJ: Lawrence Erlbaum Associates.

Gerrig, R., & Gibbs, R. W., Jr. (1988). Beyond the lexicon: Creativity in language production. *Metaphor and Symbolic Activity*, *4*,1-19.

Gibbs, R. W., Jr. (1994). The poetics of mind: Figurative thought, language, and understanding. New York: Cambridge University Press.

Gibbs, R. W., Jr. (1998). The varieties of intentions in interpersonal communication. In S. Fussell & R. Kreuz (Eds.), *Social and cognitive aspects of interpersonal communication*. (pp. 19-38). Mahwah, NJ: Erlbaum.

Gibbs, R. W., Jr., (1999). *Intentions in the experience of meaning*. New York: Cambridge University Press.

Gibbs, R. W., Jr., & Gerrig, R. (1989). How context makes metaphor comprehension seem special. *Metaphor and Symbolic Activity*, *4*, 145-158.

Grice, H. P. (1957). Meaning. *Philosophical Review*, *64*,377-388.

Haverkate, H. (1990). A speech act analysis of irony. *Journal of Pragmatics*, *14*, 77-109.

Johnson, M. (1987). *The body in mind*. Chicago: University of Chicago Press.

Kövecses, Z. (1986). *Metaphors of pride, anger, and love*. Philadelphia: Johns Benjamins.

Kövecses, Z. (1990). *The language of love*. Lewisburg, PA: Bucknell University Press.

Kreuz, R. J., Long, D., & Church, M. (1991). On being ironic: Pragmatic and mnemonic implications. *Metaphor and Symbolic Activity, 6,* 149-162.

Kreuz, R. J., & Roberts, R. M. (1995). Two cues for verbal irony: Hyperbole and the ironic tone of voice. *Metaphor and Symbolic Activity, 10,* 21-31.

Kumon-Nakamura, S., Glucksberg, S., & Brown, M. (1995). How about another piece of pie: The allusional pretense theory of discourse irony. *Journal of Experimental Psychology: General,124,* 3-21.

Lakoff, G. (1987). *Women, fire, and dangerous things.* Chicago: Chicago University Press.

Lakoff, G., & Johnson, M. (1980). *Metaphors we live by.* Chicago: Chicago University Press.

Lakoff, G., & Turner, M. (1989). *More than cool reason: A field guide to poetic metaphor.* Chicago: University of Chicago Press.

Lee, C. J., & Katz, A. N. (1998). The differential role of ridicule in sarcasm and irony. *Metaphor and Symbol, 13,* 1-15.

Leggitt, J., & Gibbs, R. W., Jr. (2000). Emotional reactions to ironic language. *Discourse Processes, 29,* 1-24.

Levinson, S. (1983). *Pragmatics.* Cambridge: Cambridge University Press.

Nayak, N., & Gibbs, R. W., Jr. (1990). Conceptual knowledge in idiom interpretation. *Journal of Experimental Psychology: General, 116,* 315-330.

Ortony, A. (1975). Why metaphors are necessary and not just nice. *Educational Theory, 25,* 45-53.

Read, S., Cesa, I., Jones, D., & Collins, N. (1990). When is the federal budget like a baby? Metaphor in political rhetoric. *Metaphor and Symbolic Activity, 5,* 125-149.

Schank, R., & Abelson, R. (1977). *Scripts, plans, goals, and understanding.* Hillsdale, NJ: Lawrence Erlbaum Associates.

Shaver, P., Schwartz, J., Kirson, D., & O'Connor, C. (1987).Emotional knowledge: Further exploration of a prototype approach. *Journal of Personality and Social Psychology, 52,* 1061-1086.

Sperber, D., & Wilson, D. (1986). *Relevance: Communication and cognition.* Oxford: Blackwell.

Turner, L., & Gibbs, R. W., Jr. (in preparation). Emotional reactions to novel and conventional metaphor.

Voss, J., Kennet, J., Wiley, J., & Engstler-Schooler, T. (1992). Experts at debates: The use of metaphor in the U.S. Senate debate on the Gulf crisis. *Metaphor and Symbolic Activity, 7,* R197-214.

Walker, N. (1990). *Feminist alternatives: Irony and fantasy in the contemporary novel by women.* Jackson: University of Mississippi Press.

Williams-Whitney, D., Mio, J., & Whitney, P. (1992). Metaphor production in creative writing. *Journal of Psycholinguistic Research, 21,*497-509.

Yu, N. (1998). *The contemporary theory of metaphor: A perspective from Chinese.* Philadelphia: John Benjamins.

– 7 –

Conflict, Coherence, and Change in Brief Psychotherapy: A Metaphor Theme Analysis

Lynne Angus and Yifaht Korman
York University

INTRODUCTION

For the past 10 years my research team and I have been exploring the role and functions of metaphoric expression (Angus, 1996; Angus & Rennie, 1988, 1989; Levitt, Korman, & Angus, 2000; Levitt, Korman, Angus, & Hardtke, 1997; Rasmussen & Angus, 1996, 1997) in the context of the psychotherapeutic dialogue. For these studies we have employed an expansive definition of metaphor (Lakoff & Johnson, 1980) in which figurative expressions are viewed as evoking a conceptual and experiential transaction between contexts of meaning. In accord with writers such as Black (1977), Richards (1936), and Turbayne (1970), this transaction between contexts is also viewed as transformative such that a new way of seeing or viewing the world is created.

Despite the specialized functions accorded metaphoric expressions, research evidence emerging from discourse analyses of therapy sessions (Angus, 1996; McMullen, 1989) suggests that the more we talk, the more we talk metaphorically. So prevalent and automatic is the recourse to metaphoric expression in verbal discourse, it quickly becomes invisible in the flow of everyday conversations.

Linguists and philosophers use the terms *frozen* or *cliché metaphors* to denote when metaphoric transactions are (mis)heard as literal descriptions by speakers and listeners alike. For example, the metaphor phrase *I feel down* is strictly speaking figurative in nature but so familiar to us in everyday English that it is instantly understood as referring to the experience of feeling depressed. As native English speakers we no longer need to carry out the imagistic and kinesthetic transposition of feelings of depression into the context of being weighed down in order to know what the speaker means when he or she uses the phrase. It is automatically understood. The term *depression* itself is metaphorical and implies a feeling of being weighed or pressed down. Numerous other metaphorical phrases associated with a pervasive feeling of sadness—*a heavy heart, carrying a heavy emotional load, it's a heavy burden*—also reflect the underlying metaphoric frame of being weighed down. It is in this manner that implicit contexts of meaning come to structure and metaphorically express—via metaphor themes—our experiences of self and others in the world.

Lakoff and Johnson (1980) argued that metaphor phrases in everyday language operate as verbal markers of metaphorically grounded conceptual systems that undergird much of our language system. These metaphor frameworks also coherently shape and resonate with our intrasubjective experiential world. It is the exploration of how metaphor themes—and by implication the underlying construct systems they represent—change in the context of successful psychotherapy, which is the focus on inquiry for this chapter. In particular, metaphor themes of interpersonal and intrapersonal conflict and depression are discussed in the context of therapy sessions drawn from two good-outcome, brief experiential therapy dyads.

METAPHOR THEMES, SELF-COHERENCE, AND THERAPEUTIC CHANGE

Starting with an early interest in metaphorical words and phrases spoken in selected therapy sessions (Angus 1992; Angus & Rennie 1988, 1989; Rasmussen & Angus 1996, 1997), we are now centrally interested in the development and evolution of metaphor themes (Angus, 1994; Levitt, Korman, & Angus, 2000; Levitt, Korman, Angus, & Hardtke, 1997) within and across therapy sessions. Specifically, we are exploring the role of metaphor themes in the creation and maintenance of coherent sense of self or selfhood for the client as well as the coherence of the intersubjective discourse co-elaborated between client and therapist in therapy sessions. Schafer (1992) suggested that metaphor themes function as implicit story-lines that give form or structure to the stories told to therapists, and underscore the connections made between the array of events described over the course of a therapeutic relationship. To this end, both McMullen (1989) and Angus (1996) have recently demonstrated persuasive evidence that the creation of elaborated metaphor themes—which span therapy sessions—is a marker of good therapeutic outcome.

Returns from an intensive analysis of metaphor themes selected from one brief, psychodynamic therapy dyad (Angus, 1994), indicate that conflict themes were clearly co-elaborated by both client and therapist and appeared to be sensitive to significant shifts in conceptions of self and others in the therapy sessions. Based on these findings, it was recommended that intensive analyses of recurring metaphor themes might be a productive method for (a) tracing the development of productive working alliances in therapy relationships, and (b) tracing the path of specific conceptual and experiential stages entailed in the successful resolution of long-standing interpersonal problems.

Additionally, utilizing qualitative research methods, Rasmussen (Rasmussen & Angus, 1996, 1997) has explored the impact of client's use of metaphor on therapist's experiences of coherence in individual therapy sessions. Analysis of therapy sessions with borderline and non-borderline psychotherapy clients indicated that borderline clients had difficulty reflexively distancing themselves from their own emotional experiences during the therapy sessions. Moreover, they often failed to respond to their therapist's attempts to metaphorically represent and organize disparate experiences into a coherent perspective or framework, during the therapy sessions. In turn, the therapists who worked with the borderline clients described feeling disorganized, misheard, incoherent, and dissatisfied with their within session interactions. In essence, the difficulty in sustaining integrating metaphor themes in the therapy sessions was highly correlated with the therapists' experiences of incoherence in, and dissatisfaction with, their therapy sessions with borderline clients.

From a constuctivist (Neimeyer & Mahoney, 1995) or post-rationalist (Guidano, 1995) perspective, the creation of self-coherence is viewed as a central processing task for all human beings. Distressing experiences that are highly discrepant with prevailing self–other conceptualizations maybe (a) not symbolized, (b) actively devalued or denied, or (c) lead to a significant reordering or change in prevailing conceptualizations. Guidano (1985) suggested that in situations of significant emotional and conceptual change, a dialectical processing of immediate self-perceptions and conscious beliefs and attitudes takes place that results in a reconstruction of the patterns of coherence that function to maintain a consistency between felt experience and self–other construals. Drawing from this theoretical framework, we argue that the generation of elaborated metaphor themes—which link and integrate disparate life experiences —are key to the development of coherence within subsystems of self and other conceptualizations in psychotherapy. And accordingly, significant self-change will be evidenced in changes in the metaphoric language clients use to express themselves to their therapists.

In accordance with the conceptualizations of Lakoff and Johnson (1980), metaphor themes are viewed as explicit verbal markers of experientially based schemes or implicit frameworks that shape and order our senses of self and others in the world. These implicit conceptual frameworks are in part shaped by cultural norms imbedded in all language systems and as such are to some extent shared by most members of a particular cultural milieu. From a constructivist perspective, however, the capacity to reflect on symbolized experiences enables human beings to make explicit and critically reflect on these implicit experiential and conceptual systems that shape the actions we take in the world as well as the meanings we attribute to the actions of others.

The symbolization of ongoing experiences in language generally—and metaphor themes specifically—enables both clients and therapists alike to reflexively evaluate and make meaning of a disparate variety of life experiences. Reflective engagement in the processing of experiential meanings, and the integration of these meanings into extant conceptual frameworks, is the ground on which psychotherapy and personal change processes are founded. Meanings and emotions that prove to be highly disjunctive with endorsed views of the self or others may evoke broad-based reevaluations of who we are and how we think and feel about ourselves. Old stories may in turn come to be understood in new ways and new metaphor themes may emerge to represent this different way of seeing and experiencing an aspect of self.

Conversely, new stories may emerge that until this point have remained unprocessed, inexplicable, and hence, not understood. Both the seeing of the known in a new or different metaphoric light, as well as the experiencing, symbolization, and assimilation of the unknown into existing metaphor themes, are viewed as key markers of successful change in psychotherapy. It is the process of metaphor theme change, in the context of good-outcome experiential psychotherapy, that is explored in this chapter.

METAPHOR THEME ANALYSES IN BRIEF EXPERIENTIAL PSYCHOTHERAPY

In order to explore the generation and elaboration of metaphor themes in psychotherapy, all sessions from two good-outcome brief experiential psychotherapies were subjected to an intensive sequential analysis for the identification of metaphor phrases and metaphor themes. The clients in this study were treated by therapists representing two different subtypes of experiential psychotherapy: client-centered therapy approach (Rogers, 1986), and process-experiential therapy approach (Greenberg, Rice, & Elliot, 1993). Client-centered is a non-directive approach to therapy where the therapist's empathic understanding, genuineness, and unconditional positive regard are encouraged. Process-experiential therapy integrates the client-centered approach with therapeutic interventions (e.g., two-chair, empty chair) drawn from gestalt psychotherapy. The process–experiential approach also emphasizes the emotional processes in people (Greenberg et al., 1993).

Three focal questions were addressed in this inquiry. First, the presence or absence of core metaphorical themes in consecutive sequential therapy sessions from two best-outcome therapy dyads was intensively analyzed and evaluated. Second, as the two therapy dyads achieved successful outcomes, and hence engendered significant therapeutic change in terms of level of depression, the question of how core metaphor themes evolved and changed over time was also explored. As little is known about how metaphor themes evolve or change during therapy, this investigation was exploratory in nature. Finally, similarities and differences between the metaphor phrases and themes in the two therapeutic relationships were also assessed.

Participants

Clients

The study includes all consecutive therapy sessions of two successful-outcome dyads. The participants were part of the National Institute of Mental Health study on depression directed by Dr. L. Greenberg of York University, Toronto, Canada (Greenberg & Watson, 1998). Potential participants, who were suffering from depression at the time, were recruited for the study. After an initial phone interview, all applicants underwent two two-hour assessment interviews in order to determine the severity of the depression. Thirty-four participants were included in the study based on the following criteria: (a) they scored more than 50 on the Global Adjustment Scale; (b) they presented a target complaint or problem issue, for example, a specific unresolved interpersonal issue; (c) they were classified on the DSM IIIR as having a major depressive episode. Excluded from the study were those individuals who were either currently in therapy, were on medication for depression, and/or identified as high risk, suicidal applicants.

The participants consented to all sessions being audiotaped, videotaped, and allowed researchers to study them. In return, participants were offered 16 to 20 sessions of free short-term psychotherapy. Therefore, participants were selected based on the assessment interviews and their willingness to consent to the conditions of the study. In order to ensure confidentiality, participants were given, and later referred to by, code. Participants were also notified that they could retract from the study at any point in time.

Selected participants were randomly assigned to one of two therapeutic approaches: client-centered or process-experiential therapy. A battery of outcome measures were completed before and after the treatment, as well as at a 6 month follow-up interval. They included: (a) Symptom Check List (SCL-90; Derogatis, 1983); (b) Beck Depression Inventory (BDI; Beck, Ward, Mendelson, Mock, & Erbaugh, 1961); (c) Inventory for Interpersonal Problems (IIP; Horowitz, Rosenberg, Baer, Ureno, & Villasenor, 1988); (d) Rosenberg Self-Esteem (RSE; Rosenberg, 1965); and (e) Target Complaints (TC; Battle, Imber, Hoehn-Saric, Stone, Nash, & Frank, 1968).

The two successful-outcome clients chosen for this study had the highest improvement rates on the outcome measures compared to the other clients in their pre-assigned therapy group. For example, the process-experiential client's

BDI score shifted from 25 before treatment to 3 at termination, whereas the client-centered client BDI score shifted from 21 to 12 over the course of the therapy relationship. Furthermore, the BDI scores remained low at 6-month follow-up. Both clients in this study were married women with two children each. The process-experiential client was in her early 30s while the client-centered client was around 60 years of age.

Therapists

Both therapists were women who were in the process of completing their PhD in Clinical Psychology. Both therapists had received extensive training in both process-experiential or client-centered psychotherapy techniques and approaches.

Procedure

In order to conduct the intensive analysis of metaphor phrases and themes, 15 sessions of the process-experiential modality and 17 sessions of the client-centered approach were audiotaped, videotaped, and later transcribed. The sessions were transcribed in accordance with the standards outlined by Mergenthaler and Stinson (1992).

Metaphor Analyses

Metaphors were identified according to Lakoff & Johnson's (1980) inclusive and encompassing definition for metaphor: "The essence of metaphor is understanding and experiencing one kind of thing in terms of another" (p. 3). Metaphor phrases initiated by either therapist or client were identified and listed. The session number from which each metaphor phrases was extracted, as well as the person who initiated the metaphor (therapist or client), were recorded. Inter-rater agreement for the identification of metaphor phrases in three therapy sessions was 87%. The two raters identified a total of 365 metaphors in the three sessions.

In order to stay "close to the text" and capture the subjective experiences of both the client and the therapist, no pre-fixed metaphor themes were assumed. The themes were inductively constructed in the following manner. The first step entailed both the identification of all metaphor phrases occurring in a given therapy session transcript and the listing of these phrases in a separate file for further analyses. The metaphor phrases selected from each dyad were analyzed on a session by session basis.

The second step involved a systematic categorical analysis of all metaphor phrases contained in a session file. Individual metaphor phrases were sorted into clusters on the basis of thematic similarities. For instance, metaphor phrases pertaining to relationships that had key words such as *stinging, struggling,* and *hard-hitting* were all clustered together under the generic theme FIGHTING. The corresponding metaphor phrases were organized such that most metaphors were sorted into clusters with only a few remaining phrases, which were classified as "unclassified metaphors." The next consecutive session was then analyzed according to the same procedure.

The identified clusters were then inductively organized according to cross cutting themes, which were termed categories. For instance, because numerous metaphor phrases were sorted into clusters such as "fighting," "winning," and "losing," clusters crosscutting the thematic category of RELATIONSHIP AS CONFLICT were identified.

In the event that a new category emerged in the analysis of later sessions, the rater reviewed the uncategorized metaphors from earlier sessions to see if they might represent the new category. Once the crosscutting themes were identified, a fellow student was provided with a list of identified metaphor phrases from two sessions in order to evaluate the agreement level of categorizing metaphors into themes. The inter-rater agreement for the categorization of metaphor phrases into the emergent themes was 82%.

RESULTS

A total of 6,064 metaphor phrases were identified in the 31 therapy sessions analyzed. Of these metaphor phrases, 3,269 were identified in the process-experiential sessions, and 2,795 were identified in the client-centered sessions. All of these metaphor phrases were included in the development of the emerging clusters and categories. Of the metaphors produced during process- experiential sessions, 85% were clustered according to themes, as opposed to 80% of the metaphors produced in client-centered sessions. The metaphors that were not clustered into themes were excluded from all further analysis. Of the metaphors that were thematically clustered, 64% of those produced in the client-centered dyad originated with the client as opposed to the 60% that originated with the client in the process-experiential dyad. Given that previous research has indicated that metaphor phrases are often interactionally co-elaborated (Angus, 1996), no distinction will be made between metaphors initiated by the therapists and those initiated by the clients. The average number of categorized metaphors per session was 199 for a process-experiential session and 129 for a client-centered session.

The thematic categorization of all the metaphor phrases resulted in the identification of 29 crosscutting themes for the process-experiential sessions and 43 themes for the client-centered sessions. The expectation that a select number of core themes would be evident both within and across therapy session was substantiated for both dyads. More than 60% of the categorized metaphors in the process-experiential dyad were classified under four themes, which included RELATIONSHIP AS CONFLICT, MARRIAGE AS A PRISON, FEELINGS ARE CONTAINED and SELF IS FROZEN VS. MELTING. The four predominant themes for the client-centered sessions— RELATIONSHIP AS CONFLICT, FEELINGS ARE CONTAINED, ANGER IS EXPLOSIVE, and SELF IS STUCK VERSUS SHIFTING—captured 42% of the categorized metaphors. The focus of this investigation was on the primary crosscutting theme, common to both dyads: RELATIONSHIP AS CONFLICT.

The conceptualization of the clients' relationships with their husbands in terms of a conflict metaphor, was the most predominant theme both within and across all the therapy sessions for both clients. In each dyad, 21% of all the metaphors produced in the sessions were categorized under the conflict theme. A quantitative analysis of the overall occurrence of metaphor phrases categorized in the global category of relationship as conflict revealed that the predominance of this theme in the therapy discourse did not change over time for either dyad.

In contrast, a more intensive analysis of the data revealed changes in the predominance of three subcategories that, taken together, comprised the "relationship as conflict theme." The first subcategory consisted of metaphor phrases involving issues of fighting and winning; for example *winning*, *battling*, *struggling*, and a *conflict* were included in this subcategory. The second subcategory was comprised of metaphor phrases involving issues of fighting but losing, and metaphor phrases such as *defense*, *withdraw*, *backing off*, *giving in*, *giving up*, and *losing* were included in this subgrouping. The third subcategory involved phrases suggesting resolution of the conflict, and phrases such as *peace*, *freedom*, *negotiation*, and *resolution* were included in this subgrouping. The metaphors classified under the theme and the different subgroupings were analyzed separately for the two therapy dyads.

Process-experiential Dyad

Overall, 569 metaphor phrases were sorted into clusters, which comprised the category RELATIONSHIP AS CONFLICT in the process-experiential dyad. An analysis of the frequencies of conflict metaphors occurring in each of the three subcategories, FIGHTING AND WINNING (44% of the phrases), FIGHTING BUT LOSING (32%), and NEGOTIATING (24%), demonstrated interesting changes across the therapy sessions.

First, early in this process-experiential therapeutic relationship the client tended to use metaphor phrases that were categorized in the metaphor theme cluster FIGHTING BUT LOSING when talking about her relationship with her husband. For example, the client said *I kind of cop out and sort of give up*, and *I am more defensive*. Consistent with the collaborative nature of metaphor themes, the therapist also used similar defensive conflict metaphor phrase when she stated *you took a risk and you just ended up being shot down.*

Beginning in session 7, however, and continuing until session 9, more metaphor phrases involved resolving the conflict rather than fighting with her husband. For example, the client said *the fighting would stop, I've resolved*, and *I feel a lot more at peace.* Unfortunately, this cease-fire attempt was short-lived as the metaphor phrases in sessions 11 through 13 shifted back to fighting but losing the battle. It was only in the last two sessions that the client shifted to predominantly discussing metaphors around fighting and winning the conflict with her husband. For example, the client used phrases like *I am strong now, I will fight it, well I do fight it*, and *I call the shots.*

Second, an important shift seems to have occurred between the 13th and 14th session. During session 14, 69% of all the conflict metaphor phrases classified were grouped in the FIGHTING AND WINNING subcategory, and the other two subcategories (FIGHTING BUT LOSING and NEGOTIATING) captured 31% of the metaphor phrases. No other conflict subcategory predominated to this extent in any other session. If the predominance of metaphor phrases clusters can be utilized as markers of critical client shifts in therapy, session 14 would be chosen for this process-experiential client. It is in this session that the client was fully able to concretize her conflictual experiences of fighting and winning the battle within her marriage. For example, the client acknowledged that *there's going to be a struggle always* and *I've accepted the struggle* but *I've overcome that. I feel secure.* She also realizes that she has strength and courage and that her husband *is weak.*

Third, the overall change from FIGHTING BUT LOSING at the beginning of therapy to FIGHTING AND WINNING at the end of therapy, was not a linear transition. It appears as though the client moved from losing the battle at the beginning of therapy to trying to resolve the conflict midway through therapy. When this strategy proved unsuccessful, the fighting with her husband resumed and the client shifted from feeling as though she was losing the conflict to winning it. The changes in the predominance of the metaphor phrase subcategories suggest that the client's first choice was to try and resolve the conflict. However, when it became evident that this goal was not achievable, she found herself back in the fighting zone and now resolved to win the fight.

Client-centered Dyad.

Overall, 569 RELATIONSHIP AS CONFLICT metaphors were identified in the client-centered dyad. An analysis of the three major subcategories of conflict, namely FIGHTING AND WINNING (44% of the phrases), FIGHTING BUT LOSING (30%) and NEGOTIATING (26%), revealed significant changes across therapy sessions.

At the beginning of therapy most of the conflict metaphors involved issues of fighting and defeat. For example, during the first session, the client described her relationship with her husband by saying *this isn't a marriage anymore, it's a power struggle*. She described their fights and her tendency to withdraw and *give in* to her husband. As therapy progressed, however, the trend slowly changed as metaphors of peace and negotiation were introduced. She described their relationship as *it's more quiet now* and *it's calm*. Toward the end of therapy, metaphor phrases expressing compromise and resolutions became more frequent, until the last session where the client stated *the negotiation has started*.

An additional trend emerging from the intensive categorical analysis of metaphor themes was the client's transformation from conceptualizing having a conflict with her husband to having a conflict with herself. In the early sessions, the client discussed the conflict and the power struggle with her husband. From session 10 on, however, and especially during sessions 13 and 14, there was an increasing tendency to refer to an internal war or conflict, within the client. For example, in session 13 the client said, *I am my own enemy*, and in session 14 she remarked *I fight with myself* and *I want peace with myself*. Thus, by the end of therapy it was understood that the "real" conflict was within the client.

In summary, the beginning of therapy was characterized by metaphors involving issues of fighting with husband and losing the battle. Toward the end of therapy, however, metaphors of negotiation and peace were introduced as well as metaphors pertaining to an internal struggle rather than a conflict with husband.

DISCUSSION

The results of the study indicate that a core set of metaphor themes predominated throughout the therapy relationships. This finding is consistent with previous studies that investigated successful therapeutic outcome clients but only employed selected or partial sessions for the metaphor analyses (Angus, 1996; McMullen, 1985). As was evident in a previous intensive analysis of metaphor themes in psychodynamic psychotherapy (Angus, 1996), it appears that the reuse and co-creation of core metaphoric themes creates a mutually understandable terminology and a context of meaning between the therapist and the client. A shared understanding between the client and the therapist is believed to be critical to the development of a working alliance. In turn, the presence of the working alliance is believed to be one of the contributing factors in facilitating a successful therapeutic outcome.

Another element, which is critical to the development of the working alliance, is the presence of a shared focus and purpose in the therapy. A shared focus for the therapy may be facilitated by consistently employing key metaphor themes as a shared framework of understanding across the therapy sessions. Because metaphors can help facilitate the identification of new conceptualizations and new solutions for problems, staying with a given metaphor theme means sustaining a few possible solutions, which in turn provides a shared focus for therapy.

Moreover, not only did a core set of metaphor themes predominate in the psychotherapy dyads, the theme "relationships as a conflict" was found to be the most predominant theme both within and across all therapy sessions, for both dyads. This finding may be explained in several ways.

One possible explanation is that both clients had similar characteristics, which might lead to a similar way of conceptualizing their experiences. Both clients were female, married with two children; both diagnosed at the onset of therapy as having a depressive episode, and both had difficulties and problematic issues with their spouses. The conceptualization of their experiences as a form of armed conflict may be grounded in the painful and protracted emotions that they felt in the context of their respective marriages.

However, others might argue that conceptualizing problematic experiences in terms of conflict metaphors is commonplace in Western culture. Theorists such as Lakoff and Johnson (1980) argued that the majority of Western cultures regard fighting and the emotional experience of conflict as an inevitable part of human life and of social interaction. From this perspective, likening a marriage to an armed conflict or war is predicated on (a) an experiential resonance to the emotional and physical pain that often occurs in conflictual situations, and (b) cultural and societal norms that view war and armed conflict as a natural and inevitable part of human life. Social constructionists on the other hand might suggest that all experiences are conceptualized and shaped by societal norms and as such it is quite predictable that conflict and fighting metaphors will be a dominant theme in conceptualizing our experiences in relationships with others (Gergen, 1985). From this frame of reference, Western culture is characterized by an emphasis on individualism, where nothing is given for free and people have to fight for their voice to be heard. In such a world, each person strives to maximize their individualistic benefits, at times at the expense of others. Within such a framework it is not surprising that people might perceive a relationship in terms of "a conflict," "a fight," or "a battle." After all, conflict and battle, as brutal as they may be, are not necessarily negative conceptualizations because they can provide a viable means to achieve peace and happiness.

Although relationship as conflict was a common and predominant theme for both clients, it is very important to note that each of them fought the "battle" in her own unique way. One client ended therapy as the winner of the conflict; the other client started negotiating peace with her husband as she realized the "real" battle was from within. This suggests that there must be more to conceptualizing experiences than culturally constructed influences, and points to the significant influence of experiential processes in the inception and articulation of change in psychotherapy.

Theme Reformulation and Dialectical Change Processes

The intensive analysis of core themes also revealed that the subcomponents of the RELATIONSHIP AS CONFLICT theme (i.e., FIGHTING AND WINNING, FIGHTING BUT LOSING and NEGOTIATING) changed and evolved as therapy progressed. For example, in the RELATIONSHIP AS CONFLICT theme the process–experiential client moved from conceptualizing and experiencing herself as losing the conflict with her husband at the beginning of therapy, to fighting and not giving in, at the end of therapy.

On the other hand, the RELATIONSHIP AS CONFLICT theme was reformulated in a different manner in the client-centered dyad as the client moved from talking about fighting the conflict with her husband, in early therapy, to negotiating and resolving the struggle, at the end of therapy. The client also moved from talking about fighting the conflict with her husband to talking about fighting the conflict with herself, a process of internalizing the conflict. These changes in metaphor phrases, themes, and subcategories in turn may map shifts in the organization of the client's experiential world.

Not only did metaphor themes evolve as the therapy progressed but they also suggested that a resolution had occurred. These findings seem to suggest that metaphor phrases and themes can reflect a therapeutic change. Siegelman (1990) suggested that in the process of psychotherapy, people search for a more coherent and encompassing reconstruction of themselves. Part of this process, she argued, entails finding new and alternative metaphors to conceptualize the self and therefore a change is likely to be evident in the metaphors clients use.

Additionally, in all metaphor phrases categories, change occurred in a nonlinear fashion. That is, clients tended to oscillate, from one session to the next, between the two extreme experiences of a given core theme. This pattern of oscillating shifts provides empirical support for Guidano's (1995) conceptualization of dialectical change in situations of significant emotional and conceptual change. Specifically, the oscillating pattern of metaphor phrase frequency may be indicative of the difficulties experienced by the client in symbolizing and concretizing the different emotions and meanings about self and others that were emerging in the therapy sessions. It is hard for clients to anticipate a linear experiential change, especially when the transition is difficult. As one of the clients in the study said, *it's like two steps forward, one step back*, when she referred to how things change in therapy.

The occurrence of patterns of dialectical or nonlinear change also has significant implications for psychotherapy process researchers. It is currently a common methodological strategy in psychotherapy research to selectively analyze a subset of sessions—often an early, a middle, and a late session—in order to exemplify significant aspects of the therapeutic relationship as a whole. However, the finding that metaphor theme subcategory change occurred in a nonlinear, dialectical fashion suggests that no one session can be taken to be representative of a phase of therapy or the entire therapy as a whole. That is, the session that researchers randomly select for investigation might be characterized as the "one step back" and might not indicate a change, even though a change had occurred.

CONCLUSIONS

The results of the study present several implications for both researchers and clinicians. First, therapists are recommended to pay attention to the metaphor phrases produced by their clients as these may convey a great deal of information about the clients' experiences. Second, the results imply that tracing metaphor themes may help in investigating the conceptual and experiential stages entailed in the successful resolution of long-standing problems. Third, the study findings suggest that metaphor phrases may reflect significant shifts in clients' experiencing and conceptualization of self and others. By tracking the reformulation of core subcategories of metaphor themes, psychotherapy practitioners and researchers can be alerted to moments of significant change for clients.

The presence of a core set of metaphor themes that are mutually shared by both the client and therapist may be used to assess the development of a productive therapeutic relationship and the presence of a working alliance. Additionally, because change in the predominance of metaphor phrase subcategories occurred in a nonlinear fashion, it may be "risky" for researchers to select only a few therapy sessions from which to search for change in metaphors themes.

Future research will explore the degree to which a metaphorical change correlates with problem resolution in post-session evaluation reports. Replicating the study in larger samples and in poor-outcome therapies will also shed more light on the role metaphors play in psychotherapy.

REFERENCES

Angus, L. E. (1992). Metaphor and the communication interaction in psychotherapy: A multidimensional approach. In S. Toukmanian & D. Rennie (Eds.), *Psychotherapeutic Change: Theory-guided and descriptive research strategies* (pp. 187-210). Newbury Park, CA: Sage.

Angus, L. E. (1996). Metaphors and the transformation of meaning in psychotherapy. In J. Mio & A. Katz (Eds.) *Metaphor: Pragmatics and applications* (pp. 71-82). Hillsdale, NJ: Lawrence Erlbaum Associates.

Angus, L. E., & Rennie, D. L. (1988). Therapist participation in metaphor generation: Collaborative and non-collaborative styles. Psychotherapy, 25(4), 552-560.

Angus, L. E., & Rennie, D. L. (1989). Envisioning the representational world: The client's experience of metaphoric expression in psychotherapy. *Psychotherapy, 26*, 372-379.

Battle, C. C., Imber, S. D., Hoehn-Saric, R., Stone, A. R., Nash, C., & Frank, J. D. (1968). Target complaints as criteria of improvement. *American Journal of Psychotherapy, 20*, 184-192.

Beck, A. T., Ward, C. H., Mendelson, M., Mock, J., & Erbaugh, J. (1961). An inventory for measuring depression. *Archives of General Psychiatry, 4*, 561-571.

Black, M. (1977). More about metaphor. *Dialectica, 31*, 431-457.

Derogatis, L. R. (1983). *SCL-90R administration, scoring and interpretation manual - II.* Towson, MD: Clinical Psychometric Research.

Gergen, K. J. (1985). The social constructionist movement in modern psychology. *American Psychologist, 40*, 266-275.

Greenberg, L., Rice, L., & Elliot, R. (1993). *Facilitating emotional change.* New York: Guilford Press.

Greenberg, L., & Watson, J. (1998). Experiential therapy of depression: Differential effects of client-centered conditions and process experiential interventions. *Psychotherapy Research, 8*, 210-224.

Guidano, V. (1995). Constructivist psychotherapy: A theoretical framework. In R. Neimeyer & M. Mahoney (Eds.), *Constructivism in psychotherapy* (pp. 93-110). Washington, DC: American Psychological Association Press.

Horowitz, L. M., Rosenberg, S. E., Baer, B. A., Ureno, G., & Villasenor, V. S. (1988). Inventory of interpersonal problems: Psychometric properties and clinical applications. *Journal of Consulting and Clinical Psychology, 56*, 885-892.

Lakoff, G., & Johnson, M. (1980). *Metaphors we live by.* Chicago: University of Chicago Press.

Levitt, H., Korman, Y., & Angus, L. (2000). "Relieving a heavy burden" : The use of metaphors of burden in good and poor outcome psychotherapy. *Counseling Psychology Quarterly, 13,* 23-36.

Levitt, H., Korman, Y., Angus, L., & Hardtke, K. (1997). Relieving a heavy burden: The experience of depression in a good and a poor outcome process experiential therapy. *Psicologia: Teoria, Investigaço e Pràtica, 1*, 329-346.

McMullen, L. M. (1985). Methods for studying the use of novel figurative language in psychotherapy. *Psychotherapy, 22*, 610-619.

McMullen, L. M. (1989). Use of figurative language in successful and unsuccessful cases of psychotherapy: Three comparisons. *Metaphor and Symbolic Activity, 4*, 203-225.

Mergenthaler, E. & Stinson, C. (1992). Psychotherapy transcription standards. *Psychotherapy Research, 2*, 125-142.

Neimeyer, R., & Mahoney, M. (1995). *Constructivism in psychotherapy.* Washingtion, DC: American Psychological Association Press.

Rasmussen, B., & Angus, L. (1996). Metaphor in psychodynamic psychotherapy with borderline and non-borderline clients: A qualitative analysis. *Psychotherapy, 33*, 521-530.

Rasmussen, B. & Angus, L (1997). Modes of interaction in psychodynamic psychotherapy with borderline and non-borderline clients: A qualitative analysis. *Journal of Analytic Social Work, 4(4)*, 53-63.

Richards, I. A. (1936). *The philosophy of rhetoric.* Oxford: Oxford University Press.

Rogers, C. (1986). Client-centered therapy. In I. Kutash & A. Wolf (Eds.) *Psychotherapists casebook* (pp. 197-208). San Francisco: Jossey-Bass.

Rosenberg, M. (1965). *Society and the adolescent image*. Princeton, NJ: Princeton University Press.
Schafer, R. (1992). *Retelling a life: Narration and dialogue in psychoanalysis*. New York: Basic Books.
Siegelman, E. Y. (1990). *Metaphor and meaning in psychotherapy*. New York: Guilford.
Turbayne, C. (1970). *The myth of metaphor*. Columbia, SC: University of South Carolina Press.

– 8 –

Conventional Metaphors
for Depression

Linda M. McMullen and John B. Conway
University of Saskatchewan

Depression: "...a noun with a bland tonality and lacking any magisterial presence, used indifferently to describe an economic decline or a rut in the ground, a true wimp of a word for such a major illness."—Styron (1990, p. 37)

In a candid and eloquent account of his own experience of depression, William Styron (1990) rued the abandonment of the term *melancholia* for its modern-day counterpart, believing that the latter simply did not do justice to the human suffering it was intended to describe. In doing so, Styron spoke to the importance of naming, not only for its ability to represent and express what a speaker intends or hopes to communicate, but for its power to influence and determine future events. He continued with his assault on the term *depression*:
"... for over seventy-five years the word has slithered innocuously through the language like a slug, leaving little trace of its intrinsic malevolence and preventing, by its very insipidity, a general awareness of the horrible intensity of the disease when out of control." (p. 37)

In this brief passage, Styron hinted at several important questions. How have the words we use to refer to what we now call *depression* changed over time? What is the historical and cultural context in which such words are not only located but thrive? In this chapter, we consider these questions by analyzing contemporary conventional metaphors for depression. We choose to focus on this particular form of language because of its capacity for illuminating questions of representation and determination and because of its centrality in human thought and action (see Lakoff & Johnson, 1980). And we reach a conclusion that is both similar to and different from that of Styron: That the dominant conventional metaphor of DEPRESSION IS DESCENT is so much a part of the fabric of our culture that it is simultaneously trite (and virtually inaudible) and associatively rich.

167

FROM MELANCHOLIA TO DEPRESSION

In what is perhaps the most comprehensive written history of melancholia and depression, Stanley Jackson (1986) traced the appearance and evolution of the terms *melancholia* and *depression* from their Greek and Latin origins. *Melancholia*, which began to appear in English writings in the 14th century, was derived from a Greek term (meaning *melaina chole*) and translated into Latin as *atra bilis* and into English as *black bile*. These translations were, of course, consistent with the presumed etiology of the disorder, which was thought to be an excess of black bile. *Depression*, which was devised originally from the Latin *de* (down from) and *premere* (to press) and *deprimere* (to press down), entered into use in English three centuries later.

The earliest entry in the Oxford English Dictionary for *depression* in its chief current sense of to bring into low spirits, cast down mentally, dispirit, deject, sadden, is credited to Robert Burton who in his 1621 publication of *The Anatomy of Melancholy* tried to look on the bright side of poverty and want by reassuring those affected by such adversities that "hope refresheth, as much as misery depresseth." According to Jackson (1986), *depression* was firmly established in discussions of melancholia by the 18th century and, as Styron alluded to, became part of modern-day diagnostic classification systems around the beginning of the 20th century. It is important to note, however, that both *depression* and *melancholia* (including variants such as melancholie and melancholy) were used and, indeed, continue to be used to refer to a disease, a condition, a syndrome, a mood, an emotion, and a variety of states of sorrow, despair, sadness, and dejection.

METAPHORS FOR MELANCHOLIA AND DEPRESSION

According to Jackson (1986), two main metaphors and one minor one have prevailed throughout the history of writings on melancholia and depression. These metaphors have clear connections to the etymologies of melancholia and depression and to the explanatory theories of the day. Dating from the time of Hippocrates, one of the earliest (and long-lasting) metaphors is that of being in a state of darkness. Clearly linked to the notion that melancholia was due to an excess of black bile, this metaphor captured not only the sense of darkness-induced fear, gloom, or dejection experienced by sufferers of melancholia, but also the clouding of thought, consciousness, and judgment that was a dominant characteristic. Authors of the time, including Galen, Alexander of Tralles, and Ishaq ibm Imran, wrote of the blackness of the bile throwing a shadow over one's thoughts and of heavy, sooty, or smoke-like vapors from thick, atrabilious blood in the stomach rising up and clouding the brain, dimming its power to apprehend and judge.

A second, main metaphor that, according to Jackson (1986), dates as far back as the notion of being in a state of darkness, is that of being weighed down or weighted down. Again, writers such as Ajax, Alexander of Tralles, and Ishaq, described minds that were weighted down, heads that contained a feeling of heaviness, and bodies that were borne down by a sensation of weight and heaviness. Jackson speculated that the bent-over head and neck and drooping posture of the melancholic person likely contributed to the aptness of these metaphors, both in the experience of the sufferer and for the observer. With the introduction and increased popularity of the term *depression* in the 17th and 18th centuries, the notion of being weighed or weighted down (or pressed down as the term denotes) continued to occupy a central place in the descriptions and accounts of dejected states.

A less dominant and long-standing metaphor is that of being slowed down. Jackson (1986) traced the introduction of this trope to the 17th century adaptations of mechanical theory, in which the body's circulatory system was thought to operate in terms of hydrodynamic principles and notions of fluid-flow. Again, the clearly observable slowing of the activity level of the depressed person contributed to the success of this metaphor, although like others, its use waxed and waned with the rise and fall of various explanatory theories. For example, Jackson reported that the language of being slowed down was used less and less with the trend away from fluid-flow theories in the latter half of the 18th century, but that we currently see a return to such language (e.g., in phrases such as slowing of thought, psychomotor retardation) in our current symptom-based diagnostic system.

All three of the metaphors described by Jackson (1986) have their origins in past centuries. How applicable are they as descriptors of what we call *depression* today?

Empirical Data

To address this question, we gleaned the metaphors used by clients in psychotherapy to describe their experience of depression to their therapists. The data came from the audiotaped psychotherapy sessions ($N = 471$) of 21 clients who participated in protocol one (therapy-as-usual) of the Vanderbilt II Psychotherapy Research Project (Henry, Schacht, & Strupp, 1990). Clients in the project received up to 25 sessions of weekly psychotherapy by experienced psychologists or psychiatrists who, for the most part, were self-described as psychodynamically oriented. There were 15 White female clients and six White male clients with predominant DSM-III (American Psychiatric Association,

1980) diagnoses of Major Depression, Dysthymic Disorder, Generalized or Atypical Anxiety Disorder, and Other or Mixed Personality Disorder. They represented a range of ages (24–62 years; M = 41 years) and educational achievement (incompletion of grade school to completion of graduate school).

The metaphors that clients used to talk about their experience of depression were selected from a large corpus of instances of figurative language that related to the major themes of therapy for a particular case, to the expression of emotion and cognitive-affective-behavioral states, and to views of self and others. We used Barlow, Kerlin, and Pollio's (1970) *Manual for Identifying Figurative Language*, in which metaphor is defined as "an implied comparison between two things of unlike nature that have something in common," (Corbett, 1965, p. 438). The selection of metaphors specific to depression was based on explicit references to depression and on culturally based knowledge of how we talk about depression.

Four main metaphors were present in our data. Consistent with Jackson's (1986) work, we found that clients spoke of DEPRESSION IS DARKNESS. Often, this metaphor took the form of likening depression to cloudy, rainy weather. For example, clients used expressions such as:

> It [last week's depression] felt pretty clouded. It was pretty cloudy. I was surprised how long it took me to clear up.

> It's really like a black cloud

> It [depression] is just like it is not even me. It's just like it is a cloud that comes over and sits over me and overcomes me and it'll stay there awhile and then it goes away.

> I have such a rainy-day attitude about everything all the time.

> I feel like the, uh, rainy day is right around the corner.

Consistent with the effects of cloudy, rainy weather, clients also spoke of feeling *dark* and of being *blue*. Inherent in these phrases is not only the sense of decreased clarity of consciousness and pervasively negative attitude and affect that are two of the hallmarks of depression, but also the not uncommon way of talking about depression as if it autonomously descends upon a person, that is, occurs independently of one's will. We return to this latter theme in subsequent sections of the chapter.

A second metaphor identified by Jackson (1986)—DEPRESSION IS WEIGHT—was also evident in our data, although there were very few instances of it. Examples included:

> It was just like I was carrying a load around [on the weekend] ... just heavy, you know, just a heavy.

> I feel just so - so heavy.

There is a sense in these examples that the entire body or the entire person, rather than a specific part of the body, such as the head (or mind) or the heart, are weighted down. Clients in our sample did not speak of their minds being "weighted down," of an experience of "heaviness in the head," or of feeling "heavy-hearted," as was noted in Jackson's account of the history of this metaphor. Instead, the experience of being burdened, and possibly of having one's freedom to move somewhat curtailed, is central.

Linked to the metaphor of DEPRESSION IS WEIGHT is another minor metaphor not mentioned in Jackson's account—that of DEPRESSION IS CAPTOR. Here, again, clients communicated a sense of restriction:

When I get depressed, I just am immobilized.

I feel trapped.

I want to break out of this.

I want to be free from it.

All three of the metaphors described so far, although clearly recognizable to most speakers of English in the Western world, accounted for less than 10% of all the instances in our data set. The most dominant and productive metaphor, which subsumed over 90% of the instances in our corpus, was what we have termed DEPRESSION IS DESCENT. As we illustrate, this metaphor is productive in the Lakoffian sense both of having a large number of linguistic expressions that code it and of carrying over many details of knowledge from the source domain to the target domain (see Lakoff, 1987, p. 384). And as we hope to argue, it is somewhat different in form and implications to the notion of being "weighed or weighted down" or of being "pressed down."

Central to the metaphor of DEPRESSION IS DESCENT were expressions about being *down* or *low*. For example, almost all clients in our sample produced statements such as:

It still takes so little for me to get depressed again, to get down.

I just was down.

I kinda went down last night and I really don't know why.

...you have the depths of depression sometimes, really feeling low.

After I saw you last Monday, I hit a low.

What would you say I should try to do when I find I'm getting into a low time?

Arising out of this central sense of depression as feeling or being "down" or "low" were numerous metaphorical entailments or elaborations derived from our knowledge and experience of descent. For example, we know that going down is quick, easy, and requires no will, but coming back up takes a long time, is difficult and effortful, and requires will. Consistent with this knowledge, clients spoke of *spiraling down, being in a down trend, having a downswing, going through a nosedive, crashing down, sliding down, sinking low, slipping into depression* and *falling in and out of depression*. These phrases denote clearly the sense that depression is conceptualized as a downward progression that, once begun, is difficult, if not impossible, to stop. In describing the difficulty of ascent after the downward progression, the clients made statements such as:

> I just have to use my own strength and my faith and, uh, to try to ... to pull myself out of it.

> I made efforts to pull myself out of that ... I made efforts to climb out.

> ...when I was so down it reminded me of ... and I was having such a terrible time pulling myself up by the bootstraps.

> I don't want to get enthusiastic, go back down, and have to slowly work my way back up.

> It's when I don't have anything there to look forward to or to go for that I have a real problem ... I get mired down.

> I didn't seem to have any get up and go ... I was kinda depressed, kinda low and I had to pick myself up and move myself along.

There is a clear sense that great personal effort, strength, and determination is required if one is to get back "up" after becoming depressed.

We also know from our experience that going down far can entail going below the surface of the earth. Clients were particularly vivid in their descriptions of the place that, to them, was depression. For example, they stated:

> I just really don't want to backslide into the dreary, dismal pit that I was wallowing around in when I first came in here.

> ...it just seemed like I was doing down into the pit and I couldn't, couldn't get out of it.

> I was at rock bottom. There was no question about it ... I ended at the bottom and there was no way up, no way to go but up.

> The bottoms are not as low. The valleys are not as deep as they used to be.

> I was at the bottom of the trough when I stated [therapy].

> This morning .. I was just way down in the gutter.

> I'm down in the dumps.

> Down in my catacombs ...

> I really got down in the depths about that.

It seems so unfair to say to [husband], "well, you can help pull me out of this slump."

I was a whole lot more in the well when [husband] told me he didn't want me to move to [different state].

Consistent with these two main elaborations were more particularized consequences. For example, a common experience is that the further one goes down, the harder it is to come back up, for example, *I got so down about that ... sometimes I just feel I just can't hardly crawl out of it*, and that if one goes down too far, one is beyond reach, as in *I spent all last weekend being real depressed— really down and unreachable*. In addition, when we are down, we know that things can fall on top of us, for example, *I just sank and said, you know, 'what's the use?', you know, just let it crumble down on top of me*, and we also know that, in most cases, once we begin to fall we must fall all the way before we can come back up, for example, *It's like I gotta reach bottom before I can begin to come back up*.

What seems to stand out in our data is that the clients in psychotherapy spoke of their experience of depression not so much in terms of feeling weighed or pressed down, but rather as a sense of movement in physical space from a higher to a lower position or place. In other words, rather than speaking about the body or the mind being weighed or weighted down, or about being pressed down by some sort of physical or psychological weight, the clients spoke predominantly of *being* or *feeling* down or low, of descent and ascent, that is, as though their person was often conceived as being at some point, position, or place on a vertical scale. One might argue that we are focusing on an ephemeral or false distinction, that, indeed, a person who reports being down or low might very well be experiencing a feeling of being weighed or pressed down. And we would not deny that this sensation of heaviness or weight is very common. However, what we are contending is that talk of being weighed or pressed down was not often explicit in our data.

"DEPRESSION IS DESCENT" AS A CULTURAL CATEGORY

How are we to make sense of the success of the DEPRESSION IS DESCENT metaphor? That is, how can we understand why it has become such a fitting metaphor? What does this metaphor tell us about our cultural model of depression? What implications does the use of this metaphor have for our experience of and reaction to depression?

One route to addressing these questions is to consider the place of orientational metaphors in our culture. As outlined by Lakoff and Johnson

(1980), these metaphors (e.g., up–down, in–out, right–left, front–back, center–periphery) structure a whole system of concepts that are fundamental to our culture in terms of spatial orientation.

Among the most common of these metaphors is up-down. Being up or ascending is associated with being happy (e.g., *She's on top of the world*), being conscious or awake (e.g., *He's an early riser*), being healthy (e.g., *She's in peak physical condition*), having control and power (e.g., *He's at the top of the ladder*), and being virtuous (e.g., *She's an upstanding citizen*). By contrast, being down or descending is associated not only with being sad or depressed but with being unconscious (e.g., *She fell into a coma*), sick (e.g., *His health is declining*), close to death (e.g., *She's very low*), dead (e.g., *She dropped dead*), subject to the control of others (e.g., *He's low man on the totem pole*), having low status (e.g., *He's at the bottom of the heap*), being poor (e.g., *She's lower class*), and lacking virtue (e.g., *That was a low-down thing to do*). In short, "up" is typically desirable and good, whereas "down" is typically undesirable and bad.

Several writers (e.g., Johnson, 1987; Lakoff & Johnson, 1980; Turner, 1991) have commented extensively on the physical/experiential basis for the up–down orientational metaphor. Specifically, they observe that as physical bodies we are usually oriented upright when we are healthy, awake, and alive; and we are often prone when we are ill, sleeping, or dead. Similarly, an erect posture is more often associated with happiness whereas a drooping posture is associated with sadness and depression. However, there are also social and cultural values and moral evaluations that are correlated with the up–down metaphor. So, in our culture, "up" is associated with high status, power, and control (which are generally considered positive and desirable) and with virtue (which is thought to be good), whereas "down" is correlated with low status and a lack of power and control (which are generally considered negative and undesirable) and with depravity (which is thought to be bad).

What do these correlations tell us about our conception of depression? Perhaps most obviously and importantly, they contribute to the sense that depression is an undesirable, unequivocally negative state, that is contains no remnants of the positive features occasionally attributed to sufferers of its predecessor, melancholia. Several scholars (e.g., Gilman, 1988; Jackson, 1986; Radden, 1987; Schiesari, 1992) have commented on this distinction between melancholia and depression. According to Jackson (1986), the association of the less severe forms of melancholia with intellectual and moral superiority, even genius, was common during the Renaissance. Because of possessing a heightened sensibility, artists, noblemen, and intellectuals of the day were thought to be especially at risk for contracting melancholia. Having a keener ability to perceive a full range of stimuli or feel a full range of emotions, gifted individuals were considered more likely to be fully aware of and affected by the environment. So, along with being able to appreciate the aesthetic side of life, they also were affected by its pathetic side and, hence, suffered from melancholia.

With its connection to nobility and genius, melancholia was also seen as a passive madness. Indeed, according to Gilman (1988), male characters in late medieval epics, who were portrayed as melancholic, were often given passive, "feminine" attributes, and the source of their madness was frequently labeled as "love sickness." However, when the origin of "love sickness" shifted over time from the mind (the highest sense) to the genitalia (which were controlled by reflexes), the social status of "love sickness" also shifted from the nobleman to the woman, from what Gilman described as "the top of [the] social ladder to the bottom" (p. 100).

Schiesari (1992) extended this analysis of the evolution and subsequent demise of melancholia and the corresponding rise of the modern-day disorder of depression by differentiating "the symbolically accredited category of melancholia and the devalued status of depression (p. 17). She proposed that when this condition was considered desirable it was associated with creativity, privilege, and men; and that when it was considered undesirable, it was stereotyped as feminine and common. According to Schiesari, "the melancholic of the past was a 'great man'; the stereotypical depressive of today is a woman" (p. 95). For some authors, then, the association between depression and that which is undesirable is clearly connected to societal values about what and who are more or less worthy.

At the heart of many social and cultural values are, of course, moral evaluations. Judgments about what is good or bad are clearly mapped onto the spatial coordinates of bodily orientation: That which is up, front, and to the right is morally positive, whereas that which is down, back, and to the left is morally negative (Hanks, 1990). The association of being "down" or depressed with that which is morally negative takes many forms. As was clear in the exemplars from our study, the place to which the clients described themselves as descending was dark, dreary, gloomy, underground, and deep—a place not unlike two abodes of the dead, the grave and hell. Indeed, sufferers of depression in previous centuries have, on occasion, been conceived as sinners, as souls deserted by God (see Jackson, 1986, p. 135).

Even in some contemporary writing, the connection between depression, the soul, and sinfulness has a prominent position. Shweder (1986), for example, proposed that the notion of "soul loss" is an idea that helps us make sense of the subjective experience of depression. Citing a feeling of emptiness as a cardinal feature of this experience, Shweder likened this feeling to the sensation of having one's soul leave one's body. Because soul loss is associated with death and sleep in most religious and cultural traditions, we find again that depression is connected, in one sense, with states of being that are often perceived as vulnerable or negative. In another sense, however, if one remains alive and awake after soul loss (as one does if one is depressed), there is a clear implication that one's spiritual connection to God has been severed. This condition of estrangement from God is clearly reflected in the overwhelming guilt and feelings of sinfulness with which depression is associated in the Judeo-Christian West (Kleinman & Good, 1985).

In addition to the connection of being "down" or depressed with that which is bad in a moral sense (i.e., sinful), being in this state is also considered bad in another sense of this word, that is, inferior or inadequate. We know, for instance, that in our culture, describing oneself as *at the bottom of the trough, down in the gutter*, and *in a slump* implies that one has, in some way, failed to achieve what is expected of one, either in terms of one's own, others', or generalized societal norms. These are the words that we use to refer to alcoholics, to the homeless, and to poorly performing athletes. In this case, low location is equated with low value and low respect. There is, then, in addition to the physical/experiential basis for the up–down orientational metaphor and its moral-evaluative correlates, a set of comparative or normative judgments that are implied by this metaphor.

As noted by Rudzka-Ostyn (1988), the vertical axis that is inherent in the up–down orientational metaphor "... can be divided into an upward and downward part only relative to some entity or location which serves as a point of reference" (p. 537). In other words, being "up" or "down" is a comparative state. For the person who is depressed, being "down" might be relative to where one perceived oneself to be the previous day, week, month, or year(s). That is, the comparison might be with oneself. Alternatively (or, perhaps, simultaneously) the comparison might be with other people (e.g., *I am down* [relative to most people I observe]).

Both Rudzka-Ostyn (1988) and Johnson (1987) have commented on the normative character of the up–down metaphor. As Rudzka-Ostyn succinctly stated, "...the location or entity relative to which upward or downward movement is measured can be correlated with any norm with respect to which one calculates increase or decrease in value" (p. 537). In the case of happy–sad (or depressed) being mapped onto "up" (the positive pole of the vertical axis) and "down" (the negative pole of the vertical axis), we see that in our culture being "up" or happy is associated with that which is desirable (or valued) and being "down" or sad or depressed with that which is undesirable (or less valued). Several writers (e.g., Lutz, 1986) have observed that in Western or Euro-American culture, being happy or displaying positive affect is considered the normal state and that the pursuit of happiness is considered a normal and basic goal. If being happy or "up" is considered the norm, then being "down" is clearly not normal or typical. Being sad, to any degree, is, then, unlikely to be easily tolerated. Or, stated another way: If being happy is the culturally normative state, then upward movement on the vertical scale implies an increase in one's already existent happy state; downward movement implies the absence of happiness.

Johnson (1987) also observed that an important aspect of scales in our culture "is the possibility of imposing numerical gradients," an aspect which "has made possible activities of measurement, quantification, and prediction that have come to define part of the distinctive character of Western civilization" (p. 123). Mapping our experience of depression onto the up–down scale entails, then, the invocation of one of our culture's most basic and defining schemas. Lakoff and Johnson (1980) similarly observed that not all cultures give the priority to the up–down orientation that we do. For example, some cultures place much more value on notions of balance or centrality.

Our particular focus on quantification and measurement supports a valuing of discreteness and comparison that is consistent with recently described conceptions of the Western self. Markus and Kitayama (1991), for example, described the typical construal of the self in Western or Euro-American culture as independent, separate, and different from others. Standing apart from others, "being one's own person," and striving to achieve and push ahead of others for one's own sake are the hallmarks of what Markus and Kitayama called the *independent view of self*. Others (e.g., Geertz, 1975; Marsella, 1985; Sampson, 1988, 1989) have also commented on the so-called Western view of the individual as a self-contained, bounded, autonomous entity, contrasting it with the non-Western (e.g., Asian, African, and Latin-American) view of the person as interdependent, unindividuated, and as striving for oneness with the environment (Bharati, 1985). Inherent in a view of self that celebrates uniqueness and separateness is the valuing of comparison and discreteness. Conceiving of depression in terms of a vertical scale invokes both the notion of discrete units and comparison, each of which is linked with our culture's conception of the salient attributes of the self.

CONSEQUENCES OF THE "DEPRESSION IS DESCENT" METAPHOR

We have tried to show that being depressed or "down" in our culture is associated with all of what we strive to avoid—for example, being "not normal," devalued, of low status, lacking in power and control, morally bad, inferior, inadequate, and less than others. We have also tried to show that the up-down orientational metaphor is a fundamental, highly conventionalized, widely used trope that is frequently employed to describe a variety of negative states. What are the possible consequences of mapping what we call *depression* onto this metaphor?

One consequence of using a highly conventionalized metaphor is that what is said may frequently go unnoticed. Just as Styron (1990) complained about the insipidity of the word *depression* resulting in a lack of awareness of the devastating effects of the disease, most native speakers of English would likely agree that describing oneself as "down" often results in, at best, a passing acknowledgment of the person's condition and, at worst, inattention. The reason for this lack of attention lies, in part, in the word's breadth of use. Just as *depression* can be used to describe not only a variety of states of sadness, such as a clinical disease, a mood, and a fleeting feeling, as well as an indentation, an economic downturn, and certain astronomical measurements, "down" is similarly used to convey various degrees of unhappiness and clinical symptoms as well as several other negative states, as observed by Lakoff and Johnson (1980) and as summarized here. It is simply used too often to refer to too many conditions or modes of being. We were struck when listening to the audiotaped therapy sessions used in our study how the therapists virtually ignored the clients' use of this metaphor and its various entailments. Even when clients used somewhat more evocative terms, such as *the gutter, the dreary, dismal pit, the bottom of the trough,* and *the well*, the therapists failed to ask for further elaboration by the clients. We can speculate that the reason for this inattention was due to an assumption on the part of the therapists that they implicitly understood what the clients meant when they used words such as *down, low, in the pits*, and *at rock bottom*. However, it can also be argued that opening up this metaphor by asking clients to elaborate might have been a way of promoting productive therapeutic discourse.

A second possible consequence of using instances of the DEPRESSION IS DESCENT metaphor stems from the metaphor's potential to elicit a large and rich network of associations, all of which are antithetical to what we strive to be in our culture. Describing oneself as "down" or "low" conveys not only sad affect, but also a host of evaluative judgments about one's adequacy and worth. In addition, using phrases such as *sliding down, sinking low, spiraling down,* and *falling into a depression* suggests not only a lowering of one's place or position but also a reduction or loss of control over where one is positioned or placed. Considering the importance we attach to our position or place in society and to the sense of having control over our lives, it is possible that many of the seemingly banal instances of the DEPRESION IS DESCENT metaphor serve a determinative function and actually contribute, in part, to the degree of misery experienced by a depressed person. That is, presenting oneself as "our culture's failure" might actually exacerbate one's level of despair.

Siegelman (1990) has written extensively about the power of metaphors to elicit thoughts, feelings, and actions. She maintains that certain crucial metaphors "..not only reflect past experience but also become filters that regulate how we see our present experience and how we project our future" (p. 65). By representing our experience of what we call *depression* in images of descent, we might be compounding the sad affect that appears to be the core of depression with associations of failure and loss of control. Speaking about ourselves as having little control over being in a disdainful position or place might easily exacerbate (or possibly even result in) a sense of helplessness and feelings of self-hatred. The arduous ascent described by clients in our sample also speaks to a vision of the future (at least in the short term) that encompasses considerable personal struggle. Taken together, these images of uncontrollable descent to and effortful ascent from a low place might serve to keep a person in this place.

Whereas this place has often been described as *a prison* (Rowe, 1978), *a jail cell* (Styron, 1990), a place where the person is alone and separated from others, in *solitary confinement* (Styron, 1990), it is also possible that being in this place might be a way for some people to opt out of the demands of 20th century life. Kleinman and Kleinman (1986) have persuasively argued that depressive illness discloses not only how an individual relates to society, but also how society affects individuals. Positioning oneself as the antithesis of what is culturally valued might allow one (however painfully) to step back from the ever-present demands in many Western societies to achieve, to push ahead, and to gain control over our surroundings (Markus & Kitayama, 1991).

CONCLUSION

Depression has been labeled the "common cold of psychiatric ills" (Kline, 1974) in Western culture, a description that attests to the frequency with which this diagnosis is given and to the run-of-the-mill nature of the state of being that it references. If depression is considered one of the primary disturbances experienced by North Americans, then the way in which we speak of it can inform us about important cultural values and imperatives. As we have tried to show, our metaphors of depression are simultaneously as common and as culturally central as this state itself appears to be.

REFERENCES

American Psychiatric Association (1980). *Diagnostic and statistical manual of mental disorders* (3rd ed.). Washington, DC: American Psychiatric Association.

Barlow, J. M., Kerlin, J. R., & Pollio, H. R. (1970) *Training manual for identifying figurative language* (Tech. Rep. 1). Knoxville, TN: University of Tennessee Press.

Bharati, A. (1985). The self in Hindu thought and action. In A. J. Marsella, G. Devos, & F. L. K. Hsu (Eds.), *Culture and self: Asian and Western perspectives* (pp. 185-230). New York: Travistock.

Burton, R. (1990). *The anatomy of melancholy.* (N. K. Kiessling, T. C. Faulkner, & R. L. Blair, Eds.) Oxford, England: Clarendon Press. (Original work published 1621)

Corbett, E. P. J. (1965). *Classic rhetoric for the modern student.* New York: Oxford University Press.

Geertz, C. (1975). On the nature of anthropological understanding. *American Scientist, 63,* 47-53.

Gilman, S. L. (1988). *Disease and representation: Images of illness from madness to AIDS.* Ithaca, NY: Cornell University Press.

Hanks, W. F. (1990). *Referential practice: Language and lived space among the Maya.* Chicago: University of Chicago Press.

Henry, W. P., Schacht, T. E., & Strupp, H. H. (1990). Patient and therapist introject, interpersonal process, and differential psychotherapy outcome. *Journal of Consulting and Clinical Psychology, 58,* 568-774.

Jackson, S. W. (1986) *Melancholia and depression: From Hippocratic times to modern times.* New Haven, CT: Yale University Press.

Johnson, M. (1987). *The body in the mind: The bodily basis of meaning, imagination, and reason.* Chicago: University of Chicago Press.

Kleinman, A., & Good, B. (Eds.). (1985). *Culture and depression: Studies in the anthropology and cross-cultural psychiatry of affect and disorder.* Berkeley, CA: University of California Press.

Kleinman, A., & Kleinman, J. (1986). Somatization: The interconnections in Chinese society among culture, depressive experiences, and the meanings of pain. In R. Harré (Ed.), *The social construction of emotions* (pp. 429-490). Oxford: Basil Blackwell.

Kline, N. S. (1974). *From sad to glad: Kline on depression.* New York: Putnam.

Lakoff, G. (1987). *Women, fire and dangerous things: What categories reveal about the mind.* Chicago: University of Chicago Press.

Lakoff, G., & Johnson, M. (1980). *Metaphors we live by.* Chicago: University of Chicago Press.

Lutz, C. (1986). Depression and the translation of emotional worlds. In R. Harré (Ed.), *The social construction of emotions* (pp. 63-100). Oxford: Basil Blackwell.

Markus, H. R., & Kitayama, S. (1991). Culture and the self: Implications for cognition, emotion, and motivation. *Psychological Review, 98,* 224-253.

Marsella, A. J. (1985). Culture, self, and mental disorder. In A. J. Marsella, G. Devos, & F. L. K. Hsu (Eds.), *Culture and self: Asian and Western perspectives* (pp. 281-308). New York: Travistock.

Radden, J. (1987). Melancholy and melancholia. In D. M. Levin (Eds.), *Pathologies of the modern self: Postmodern studies on narcissism, schizophrenia, and depression* (pp. 231-250). New York: New York University Press.

Rowe, D. (1978). *The experience of depression.* Chichester, England: John Wiley & Sons.

Rudzka-Ostyn, B. (1988). Semantic extensions into the domain of verbal communication. In B. Rudzka-Ostyn (Ed.), *Topics in cognitive linguistics* (pp. 507-553). Amsterdam: John Benjamins.

Sampson, E. E. (1988). The debate on individualism: Indigenous psychologies of the individual and their role in personal and societal functioning. *American Psychologist, 43,* 15-22.

Sampson, E. E. (1989). The challenge of social change for psychology: Globalization and psychology's theory of the person. *American Psychologist, 44,* 914-921.

Schiesari, J. (1992). *The gendering of melancholia: Feminism, psychoanalysis, and the symbolics of loss in Renaissance literature.* Ithaca, NY: Cornell University Press.

Shweder, R. (1986). Menstrual pollution, soul loss, and the comparative study of emotions. In R. Harré, (Ed.) *The social construction of emotions* (pp. 182-215). Oxford: Basil Blackwell.

Siegelman, E Y. (1990). *Metaphor and meaning in psychotherapy.* New York: Guilford.

Styron, W. (1990). *Darkness visible: A memoir of madness.* New York: Random House.

Turner, M. (1991). *Reading minds: The study of English in the age of cognitive science.* Princeton, NJ: Princeton University Press.

PART III

Social and Cultural Dimensions

– 9 –

Emotion, Verbal Expression, and the Social Sharing of Emotion

Bernard Rimé, Susanna Corsini,
Gwénola Herbette
University of Louvain at Louvain-la-Neuve

INTRODUCTION: EMOTION AND THE SHARING OF EMOTION

The following scene was observed by one of us in a city of southern Italy. A traffic accident had just occurred at a street intersection. A car had hit two youngsters riding a motorbike. The two severely wounded victims were lying down on the street. In an instant, the crowd of gaping people around them was so thick that the ambulance had to stop at some distance and the medics could hardly get through. Particularly striking was the following fact: Most of the people in the crowd were using their cellular phone, reporting on-line to some close person the emotional scene they were witnessing. In this anecdotal observation, people who experienced emotion evidenced a marked need to share it and to talk about it. Before the advent of cellular phones, in such a situation, this need would have been expressed on-site: The individuals gaping at an accident scene would have been observed to talk to each other; the observer would have missed the fact that later on, when back home, all these people would have talked again about the scene and told their intimates what they had witnessed. Now that modern technology makes it possible, cellular phones clearly revealed that people who have the urge to share an emotion do so with their intimates rather than with unknown individuals around them.

In commenting this anecdote, we thus assumed that "people who experience emotion evidence a marked need to share this emotion and to talk about it." Do we need to be in southern Italy to observe such a phenomenon? Is it particularly manifested among people who are under the shock of witnessing an accident scene? Actually, the answer to both questions is negative. The urge to share an emotional experience by talking about it is a very general manifestation. It is elicited as soon as an emotion is experienced, whatever the type of this emotion. It was observed in every culture in which it was investigated.

In this chapter, we review the research we conducted in the last decade on what we called "the social sharing of emotion." Under this label, we examined a process that takes place during the hours, days, and even weeks and months following an emotional episode. It involves the evocation of an emotion in a socially shared language to some addressee by the person who experienced it. Very generally, this person will talk with others about the event's emotional circumstances and his/her feelings and reactions to it. In particular cases, the addressee is only present at the symbolic level, as is the case when people write letters or diaries (Rimé, 1989; Rimé, Mesquita, Philippot, & Boca, 1991a). We first review the basic evidence in support of the view that, very generally, emotion is followed by the social sharing of emotion.

EMOTION AND THE SOCIAL SHARING OF EMOTION: BASIC SUPPORTING EVIDENCE

Evidence From Autobiographic Data

The first studies that tested this prediction relied on collecting autobiographic data. Participants were instructed to recall and briefly describe an emotional experience from their recent past corresponding to a specified basic emotion (e.g., joy, anger, fear, shame, or sadness). They then answered questions about their sharing of this episode: Did they talk about the episode with others? With whom? How long after the emotion? How often? (Mesquita, 1993; Rimé, Mesquita, Philippot, & Boca, 1991; Rimé, Noël & Philippot, 1991; Vergara, 1993). Eight independent studies of this type were reviewed by Rimé, Philippot, Boca, and Mesquita (1992). They involved 1,384 emotional episodes reported by 913 respondents ranging in age from 12 to 72 years. The data showed that, according to the study, 88% to 96% of the emotional experiences gathered were socially shared. These proportions were independent of age and gender. Thus, contrarily to widespread stereotypes, it did not seem that women were more prone than men to share their emotions.

Collected data included samples of Belgian, southern French, Dutch, Surinamese, Spanish Basque, and Italian respondents. Comparison of these various cultures failed to evidence differences with regard to the rate of social sharing. Moreover, neither the type of basic emotion (fear, anger, joy, sadness) nor the valence of the emotional experience (positive or negative) elicited differences in the proportion of shared episodes. The modal pattern for an emotion sharing was to be initiated early after the episode. Participants indicated that they first shared the emotional event on the same day it happened in about 60% of the cases across studies. In addition, participants talked about the event "several times" with "several persons," showing that emotion sharing is a repetitive process that involves several recipients. Extent of sharing (i.e., number of repetitions and number of recipients) was positively correlated to the intensity of the emotional experience (r =. 49, p<.001; [Rimé, Mesquita, et al., 1991]): The more intense the emotion felt by the participant, the more he/she talked about the event.

Thus these initial sets of data strongly supported the view that "every emotion tends to be socially shared." Obviously, however, the ex post facto nature of these data limited the evidence. The possibility existed that respondents selectively remembered episodes that they did socially share. Thus, autobiographic data are at risk of being unduly inflated. Also, they are open to reconstructive biases.

Evidence From Diary Data

In order to overcome these methodological limitations, the time elapsed between the emotional episode and the investigation of social sharing manifestations related to this episode should be minimal. To achieve this goal, diary methods were attempted. For periods ranging from 2 to 3 weeks, participants completed a questionnaire every night before going to bed. First, they briefly described the event that had affected them most that day. Then, they rated this event on emotional intensity scales and on a basic emotions check list. Finally, they answered a number of questions about the chosen event, including questions concerning social sharing. In this manner, the risks of selective and reconstructive biases were considerably reduced. Moreover, the diary method had the advantage of allowing the observation of social sharing elicited by everyday life emotional states of low and moderate intensities.

Several studies of this type were conducted (Rimé, Philippot, Finkenauer, Legast, Moorkens, & Tornqvist, 1994). The autobiographic studies had showed very consistently that an emotional experience is shared for the first time during the day it happened in about 60% of the cases. In order to support observations from these studies, the diary method should thus confirm that daily emotional episodes are shared before the next night in an average of 60% of the cases. The diary data were exactly in line with this prediction. To illustrate, in one of the diary studies, emotional episodes were collected from 34 participants during 14 consecutive nights (Rimé et al., 1994, study 2). The episodes were classified according to the type of primary emotion involved: joy, surprise, fear, anger, shame, contempt, and so on. On average, 57.9% of the 461 events reported were shared the day they happened. This result did not vary as a function of specific emotions. However, a notable exception was that a marked trend indicated less sharing for the rare cases of shame.

To conclude, findings from diary studies in which the interval between the episode and its recall is at maximum one day replicated those from autobiographic studies involving intervals of weeks, months, or even years. Thus, it does not seem that findings from autobiographic studies can be explained by selective or by reconstructive memory biases: Social sharing behaviors appear to be a usual consequence of exposure to emotion.

Evidence From Follow-up Data

To further test this conclusion, a third method was adopted. We called it the "follow-up procedure." Having the investigators "pre-select" a target event prevented any selection bias by respondents. Participants were contacted at the time of their exposure to some important emotional event and were then followed up for several weeks. One such study was conducted in an emergency care clinic, with 39 victims of traffic, domestic, or work accidents (Boca, Rimé, & Arcuri, 1992). Participants were contacted when first receiving treatment at the hospital and were re-contacted 6 weeks later. At this follow-up session, they answered questions about accident-related social sharing. In another follow-up study, the target emotional event was child delivery as experienced by 31 young mothers (Rimé et al., 1994, study 4). When leaving the maternity ward, they were given social sharing questionnaires to be completed during each of the 5 subsequent weeks.

Five other follow-up studies were conducted using this procedure. The target emotional episodes involved: (a) bereavement; (b) academic examination; (c) first blood donation; (d) attending the dissection of a human corpse; (e) performing first dissection (for a review, see Rimé, Finkenauer, Luminet, Zech, Philippot, 1998). In each of these studies, follow-up questionnaires assessed the occurrence of social sharing immediately after the event and again at various intervals. All the follow-up studies offered evidence that was perfectly consistent with earlier findings.

In the first study, involving accident victims, participants shared the accident episode in 93% of the cases. They did so in the course of the first day (60%). They shared the episode in a repetitive manner (i.e., more than twice) for 83%, and about every day for no less than 35% of the cases (Boca et al., 1992). Findings from the child delivery study indicated that, during the first and second weeks after their release from the maternity ward, the percentage of women who spoke about the delivery experience was extremely high, 97% and 90% respectively. These figures decreased progressively during the three subsequent weeks, with the delivery being shared in 55%, 52%, and 32% of the cases respectively (Rimé et al., 1998).

Follow-ups from the five other studies indicated that, independent of the type of emotional episode, social sharing occurred during the week following the episode at rates closely matching those in the autobiographic studies (i.e., around

90% of the cases across studies). With the exception of blood donation, which is a relatively minor emotional episode, the proportion of episodes that still elicited sharing during the second week was virtually the same as during the first week. Marked decreases were then generally observed in the following weeks or months. Steeper extinction slopes were generally observed for less intense emotional events (Rimé et al., 1998).

Altogether, these findings, in which selective memory biases were precluded, confirmed the data of the autobiographic studies: When people are exposed to emotion, whether positive or negative, they usually engage in social sharing.

Evidence From Experimental Data

As this proposition clearly involves a causal relation linking emotion and social sharing, observational techniques could not really suffice in testing it. Experimental tests had to be conducted. We thus designed several laboratory studies in which emotion was induced by exposing volunteers to movie excerpts (Luminet, Bouts, Delie, Manstead, & Rimé, 2000). Volunteer students participated together with a same-sex friend in a so-called "cooperation study." On arrival at the laboratory, one member of each pair was randomly assigned to one of three emotion-inducing conditions (i.e., high, moderate, or low emotional film), whereas the other participant was instructed to complete an irrelevant task in another room. Ratings by participants exposed to the movie confirmed that the three films were significantly different for intensity of negative emotions, whereas the induced emotions were comparable. After the movie, which lasted 3 minutes, the target-participant and his or her friend were brought together in a waiting room and left alone for 5 minutes. Their conversation was unobtrusively tape recorded. Independent judges later rated the recordings for time talked about the movie and proportion of words referring to the movie. In a second experiment, using the same design and movie conditions, additional measures were used to assess induced emotions. The target participant's face while viewing the movie was covertly video recorded and later rated for intensity of facial expression and duration of gaze aversion. A third experiment was conducted to control for a number of alternative explanations and to verify in how far the effect of exposure to the movie would generalize beyond the laboratory situation.

Results from the first experiment revealed that, compared to participants in the two other conditions, those exposed to the highly emotional movie talked more about their experience. To illustrate, nearly 40% of the words spoken by these participants referred to the movie, as compared to less than 5% of the words spoken by those in the other two conditions. It thus seems that viewing a 3-minute emotional movie is sufficient to elicit a process of social sharing. Unexpectedly, however, the moderately intense movie failed to elicit more social sharing than the non-emotional control movie. The second study exactly replicated the results of the first one. First, the three conditions were significantly different for intensity of negative emotions. Second, concerning the amount of social sharing, the high emotional movie differed significantly from both the low and the moderate emotional movie. Again, the latter two conditions failed to differ from each other.

In a supplemental analysis, individual differences in intensity of emotional reactions to the movies evidenced marked correlations with the indices of extent of social sharing taken from the post-movie conversations. In other words, the more participants were emotionally aroused by the movies, the more they talked about it with their friend in the waiting room situation that followed. The same pattern of results was found for intensity of facial expression and duration of gaze aversion: Both the low and the moderate conditions differed significantly from the strong condition, but not from each other. These results perfectly matched those observed for social sharing. Overall, these two laboratory investigations offered support for the general notion that emotions elicit social sharing. In the third experimental study, the ecological validity of the two previous sets of results could be established. Participants' ratings of their sharing during the 2 days following the experiment paralleled the waiting room findings: The low and moderate intensity movies did not differ from each other, but both differed from the high intensity one.

These experimental investigations demonstrated that being exposed to an emotion elicits the social sharing of this emotion, thus confirming autobiographic, diary, and follow-up findings. However, the lack of difference in the effects of the low and moderate emotional intensity conditions opens the possibility that social sharing occurs only when an intensity threshold is exceeded.

Emotional Secrets

The studies reviewed above consistently showed that the proportion of emotional episodes that are socially shared is very high (up to about 90%). However, about 10% of emotions are not socially shared. The question arises of what the characteristics of these emotional events are that prevent them from being shared. Finkenauer and Rimé (1998) conducted two studies designed to determine the eliciting circumstances of secrecy by investigating the differences between emotional events that are socially shared and those that are kept secret. Investigating matters that people do not want to talk about represents a challenge. Yet, it was assumed that participants would actually be willing to provide information about their secret if two conditions were met. First, absolute anonymity should be guaranteed. Second, participants should in no way be asked to reveal their secret. By adopting procedures respecting these two conditions, high participation rates were obtained.

In the first study, students recalled both a shared and a non-shared emotional event and rated each event on various dimensions of the emotional experience. Results showed that neither the intensity of the emotion felt when the event occurred nor the intensity of the emotion still felt at the time of responding discriminated shared from non-shared episodes. Also, non-shared emotional experiences were no more or no less negatively valenced than shared ones. A second study was conducted on a large sample of respondents whose age ranged from 16 to 70 years. In this study, a between-subject design was used and additional measures were included to assess event-related stress and traumatic impact. None of the results supported the hypothesis that non-shared events are more traumatic than shared ones. As in the first study, both types of events were equally negative. Overall, these results suggest that high intensity of emotion or trauma is not a necessary precursor to emotional secrecy.

If emotional intensity and valence do not predict emotional secrecy, are there other characteristics of the emotional experience that do? The two studies on emotional secrecy examined this question. Both revealed that non-shared emotional episodes elicited more intense feelings of shame and guilt than shared ones. Also, emotional appraisal ratings revealed that emotional experiences kept secret involved greater personal responsibility for the event than shared experiences. Finally, data showed that non-shared experiences were initially associated with attempts to hide one's feelings or emotions, an action tendency typical of ashamed persons (e.g., Tangney, 1991).

To conclude, social sharing is the rule after an emotion and secrecy is the exception. Some characteristics of an emotion such as the subjective experience of shame or of guilt appear to play a central role in eliciting secrecy.

INDIVIDUAL AND CULTURAL ASPECTS

We now turn to the question of the generality of the social sharing of emotion. How far does the principle that "every emotion tends to be socially shared" holds across age groups? Are there important individual difference variables that affect the propensity to share emotions? How far does the principle hold in non-Western cultures?

Sharing of Emotion in Children

Two studies were conducted in order to investigate the social sharing of emotion among children (Rimé, Dozier, Vandenplas, & Declercq, 1996). They addressed very basic questions: (a) Do children evidence social sharing after exposure to emotion? (b) If yes, does their sharing evolve with age? (c) If children do share their emotions, who are their sharing targets? Peers, parents, siblings, others?

In the first study, participants were male and female school children from two age pools, either 6 or 8 years old. Each child was individually told a narrative by a female experimenter. Half of the children were randomly assigned to either a high or a low emotion story. The high emotion story was "Xandi and the monster," a narrative especially developed by previous research in order to both induce fear among children and to show them ways of coping with fear. The low emotion story described in an interesting and attractive manner "life at a farm." Immediately after the end of the narrative, each child was brought to a playroom in which children of the same age were busy playing. During a 15-minute period, the children were observed and social sharing behaviors were monitored. It was found that social sharing occurred only very infrequently among the various children, with no significant difference between emotion conditions, nor between age groups. This suggested either that children in this age range do not yet socially share their emotions, or that children of the same age are not appropriate social sharing targets.

Further data argued in favor of the latter alternative. Parents of each child involved in the study were asked to complete a questionnaire on their child's behavior during the evening after they were exposed to the narrative. These ratings revealed that 71% of the children who heard the high emotion narrative shared it with their parents, as did 42% for those who heard the low emotion narrative. The difference was statistically significant, thus evidencing a higher occurrence of social sharing after exposure to a more emotional condition. It should be noted that the social sharing rate of 71% recorded on the same day the emotion happened is quite consistent with the observations on adults showing that roughly 60% of emotional experiences are shared during the day they happened. As is the case among adults, social sharing tended to be more repetitive among children who heard the more emotional narrative. Finally, as compared to children who were exposed to "life at a farm," children who heard the more emotional narrative were also rated by their parents as having expressed their feelings more while sharing. It thus seems that social sharing occurs among children in a manner that does not basically differ from what is observed among adults in this regard.

However, it did not seem that children shared with listeners other than parents. Sharing with other people was reported by parents in only 12% of cases for children in the high emotion condition, and in 4% for those in the low emotion condition, with no significant difference between the two. It should provisionally be concluded that among children, parents tend to be the exclusive target for the sharing of emotions.

A second study was conducted among boys aged 8 to 12 who attended a scout camp. In this context, they took part in a traditional "night game" in which the kidnapping of one of them was simulated. This type of game is very popular in the country in which the study was conducted and has been practiced by many generations. It is usually run every year, so most of the boys know about it. Nevertheless, every time it occurs, children "play the game" and accordingly do feel emotionally involved.

Immediately after the game, each child was asked to report his game-related emotionality on a 20-degree "emotion thermometer." A group average of 9 was recorded, suggesting that a moderate intensity emotion was induced. They all went back home the next day. Three days after the game, the children's parents were asked to complete a brief questionnaire on their child's sharing behaviors since the camp. These ratings showed that children shared the night game in 97% of the cases. Sharing was recurrent in most cases. Six days after the game, parents were asked to complete another questionnaire concerning the sharing that occurred since the previous one. Sharing rates amounted to 39%, thus showing a sharp decline over time.

One week after the game, the children themselves were asked to complete a social sharing questionnaire. Social sharing rates recorded from these data amounted to 87% of the sample. Children also reported their sharing to be recurrent, with a modal answer approaching three repetitions. They were also asked with whom they shared. As was the case in the first study, parents clearly emerged as the privileged sharing partner in this age group. Indeed, 93% of the children reported that they shared the episode with their father and 89% that they shared with their mother. Much lower figures were recorded for other members of the family circle, with 48% of the children sharing with their brothers or sisters. Finally, other peers were mentioned by 37%, best friend by 33%, and grandparents by only 5%.

To conclude, the data issued from these two studies indicate that social sharing of emotion is a part of children's natural repertoire, at least for the investigated range of ages—6 to 12 years old. They also show that parents are the initial social sharing targets.

Social Sharing of Emotion Among Elderly

Old age is often described as a time when emotions diminish in intensity and their expression becomes rigid (Banham, 1951). Old age is considered to be characterized by affective quiescence (Cumming & Henry, 1961). This would lead to the prediction that the process of sharing emotions would progressively vanish with age. However, in sharp contrast with traditional perspectives, recent theoretical advances have proposed that age is associated with positive emotional development. To illustrate, Labouvie-Vief and colleagues (e.g. Labouvie-Vief & Blanchard, 1982; Labouvie-Vief & Devoe, 1991) stressed that greater age involves higher levels of ego development. As a consequence, older persons show greater emotional complexity, enhanced self-regulation of emotion, and a superior understanding of emotions when compared with younger persons. A comparable accent on the role of emotional life in older age is found in socioemotional selectivity theory (Carstensen, 1987, 1991, 1993). This theory proposes that as people approach the end of their life, emotional quality of social encounters becomes more important than the acquisition of information or basic survival functions (e.g., reproduction) which predominated before. Thus, the two theoretical views consider that with older age, people become "experts" on emotions. This leads us to expect that sharing of emotion does not vanish among elderly. On the contrary, older persons should share even more than younger persons.

A diary study was conducted in order to test this prediction (Rimé, Finkenauer, & Sevrin, 1995). Older adults (60–75 years) and elderly adults (76–94 years) were compared to a group of younger adults (25–40 years). Participants completed a questionnaire on the most emotional event of the day for five successive evenings. Items assessed emotional feelings and responses as well as the sharing of the emotional event. The proportion of emotional events that were shared the day of their occurrence amounted to 64% among younger adults, a figure close to the 60% found among students in former diary studies. Confirming the second prediction, the corresponding figures were significantly higher among elderly respondents, with 77% among the older adults and 85% among the elderly adults.

Thus, rate of social sharing increased with age. This was also the case for the number of times that each event was shared. These findings are not easily reconciled with traditional stereotypes stressing the poverty of affective life in the elderly. Conversely, they fit the recent theoretical views proposed by Labouvie-Vief (e.g., Labouvie-Vief & Blanchard, 1982; Labouvie-Vief & Devoe, 1991) and by Carstensen (1987, 1991, 1993). It is possible that the three groups differ on other variables that mediate the observed relationship between extent of social sharing and age (e.g., time to share, availability of sharing targets, mobility). Future studies should aim to examine the antecedents, consequences, and mediating variables of the increased sharing observed among the two older age groups.

Personality Variables and the Social Sharing of Emotion

How far is the process of social sharing of emotion accounted for by individual differences? To answer this question, the relationship between some personality dimensions and social sharing was investigated. Specifically, the "Big Five" factors (Neuroticism, Extraversion, Openness, Agreeableness, and Conscientiousness) (e.g., Digman, 1990; Goldberg, 1990) and a more specific personality dimension called "alexithymia" were hypothesized to be directly related to the social sharing of emotion. The alexithymia concept (Sifneos, 1973) addresses traits supposed to characterize psychosomatic patients. It involves (a) a difficulty in verbalizing emotions, and (b) constricted imaginative processes (Taylor, Bagby, & Parker, 1997).

Two studies conducted by Luminet, Zech, Rimé, and Wagner (2000) examined this relationship. In the first one, 99 French-speaking Belgian undergraduate students were asked to report the most negative emotional episode they had experienced during the three previous months. They gave a full account of the event, described their emotional reactions, and rated the extent and content of event-related social sharing. Subsequently, they completed two personality questionnaires: (a) the NEO PI-R, which covers the Big Five factors, and (b) the Bermond-Vorst Alexithymia Questionnaire, which assesses among others dimensions "Poor verbalization" and "Poor fantasy life." Hierarchical regression analyses revealed that none of the Big Five dimensions was related to extent of sharing. Consistent with the assumption underlying the concept of alexithymia, Poor verbalization was negatively related to extent of social sharing. Poor fantasy life was found to be unrelated.

A second study was conducted on a sample of 101 British students. This time, participants were asked to recall both a recent positive and a recent negative emotional episode from their life and to rate their emotional reactions and social sharing for both. They also answered the NEO PI-R and the Alexithymia questionnaires. With regard to negative events, again none of the NEO PI-R dimensions was associated with extent of sharing. Poor verbalization was again negatively related to extent of sharing. This time, the second alexithymia dimension, Poor fantasy life, was also negatively related to extent of sharing. For positive emotional episodes, none of the seven examined dimensions predicted extent of sharing.

To conclude, general personality dimensions such as the Big Five have no obvious predictive value for the social sharing of emotion. However, alexithymia, a very specific personality dimension linked to verbal expression of emotion, was shown to be a consistent predictor of social sharing, at least for negative events.

Sharing of Emotion and Culture

Data from countries differing in language and/or culture—Belgium, the Netherlands, southern France, the Basque country in Spain—were already examined by Rimé et al. (1992). Because the samples were comparable for rate, frequency, and delay of sharing, it was concluded that sharing of emotion is a widespread phenomenon. Comparing Dutch, Surinamese, and Turkish respondents living in the Netherlands, Mesquita (1993) also observed very high rates of sharing in each sample but found differences in shared content. This leads to the hypothesis that sharing is common among human beings, but that sharing modalities vary across cultures. Yet, all the available evidence has been limited to people living in Western European countries.

Rimé, Yogo, and Pennebaker (1996) collected data from students of social sciences or humanities in six different locations: four in Asia (Korea, Singapore, and two different locations in Japan) and two in the Western world (France and United States, Texas). Participants completed a brief recall questionnaire. They first recalled their most recent, important, unpleasant emotion and then answered items assessing the experience and its sharing. The samples did not vary in any important respect for intensity of reported emotion. Results showed that emotions were shared by a high proportion of people in all six locations (from 78% to 96% of the cases). This confirms the cross-cultural generality of the phenomenon.

Yet, samples from some of the Asian locations generally reported slightly lower rates of emotion sharing than did samples from Western locations. To illustrate, in the Korean sample, more than 20% of the emotional episodes were reported as "never shared" whereas in the American sample, this was the case for only 5%. Further cross-cultural variations emerged for sharing modalities. Sharing was repetitive in every culture, yet the number of repetitions was markedly higher in Western samples (5 times) than in Asian ones (2-3 times). The delay between the emotional event and its first sharing also varied across cultures and again contrasted Asian and Western samples, with shorter delays for the latter (1-2 days) and a particularly long delay for the Singapore sample (4-5 days).

Data on sharing targets further revealed important cross-cultural generalities and cross-cultural variations. In all six samples, best friends were mentioned with equal frequency and were by far the most important sharing targets. Strangers were rarely mentioned in the six groups. With respect to family members, marked contrasts existed between Asian and Western cultures. Indeed, the former reported less sharing with their spouse or partner, parents, and siblings than the latter. In all six groups, sharing only rarely involved grandparents. Yet, they were more often referred to by both the American and French samples than by the Asian ones. The data thus reliably suggest that, in this age group at least, Asian and Western cultures do contrast in their degree of inclusion of the nuclear family members in the social sharing of their emotions.

Yogo and Onoe (1998) examined the social sharing of various basic emotions in a large sample of Japanese students using an autobiographic data collection procedure. For every type of basic emotion, sharing was observed in a very high proportion of the episodes: in 98% of the cases for episodes of happiness, in 96% for fear, in 94% for anxiety, in 89% for love, in 88% for sadness, in 87% for disgust, and in 83% for anger. As was the case in Western data, episodes of guilt and of shame were shared in somewhat lower proportions, respectively 79% and 74%.

Singh-Manoux and Finkenauer (2000) compared the social sharing of three basic emotions among Indian adolescents living in India, Indian adolescents living in England, and English adolescents living in their home country. Across samples, the proportion of shared episodes was somewhat lower than in the other studies: 77% of all reported emotions (N = 555), with figures of 80% for fear, 82% for sadness, and 69% for shame. Respondents shared their experiences 2.6 times on average, and with an average of 2.5 people. No significant differences were observed between the three cultures for these various results. Yet, detailed analyses of variables such as latency of sharing or type of sharing target revealed a good deal of cultural variations suggesting that sharing patterns are not unique.

These data allow us to conclude that social sharing of emotion is not limited to Western cultures. It is observed in the Asian world as well. Yet, although they were based on a very brief questionnaire, the collected data revealed an abundance of cultural differences in sharing modalities, suggesting that this is a particularly promising avenue for future investigation.

SOCIAL ASPECTS OF SOCIAL SHARING

Up to here, we have adopted an essentially individualistic perspective on the social sharing of emotion. Yet, this is a fundamentally social process: Other persons are told emotional experiences. Who are these people whom we select in order to let them know about the important events that happen to us? In addition, these people will probably not remain unaffected by what they are told. How do they react? How do they behave with regard to the person who shares the emotion? Do they experience emotion in turn? Once the sharing interaction is over, does it have some later impact on the recipient of the sharing?

Targets of Social Sharing

Who are the addressees in the process of sharing an emotion? An interesting trend emerged from research considering the various age groups. As mentioned earlier, Rimé, Dozier, et al. (1996) exposed children aged 6 or 8 to an emotion-eliciting narrative and monitored the occurrence of sharing behaviors in the following hours. Virtually no sharing toward peers of the same age was observed. Most children shared the emotional episode with their father and mother when back home in the evening. Sharing with other family members was exceptional. It thus seems that among children of this age, attachment figures alone are taken as social sharing targets. In the next age group—8 to 12—some evolution is already apparent in this regard. Preadolescents participating in a summer camp were followed up after an emotion-eliciting night game. Attachment figures obviously still play the major role in the social sharing of emotion at this age, as 93% of the preadolescents shared this experience with their mother and 89% with their father. However, the range of targets was now extended to siblings and, to some extent, to peers. About half of the preadolescents talked to their siblings, and about one third shared their experience with some other participant of the summer camp.

The question of the type of partner chosen for the social sharing process was investigated in other age groups: adolescents (aged 12–17), young adults (aged 18–33), mature adults (aged 40–60) (Rimé, Mesquita, et al. 1991; Rimé, Noël, et al., 1991; Rimé et al., 1992). It was observed that for people of all ages, targets of social sharing are confined to the circle of intimates (i.e., parents, brothers, sisters, friends, or spouse/partner). People who are not part of this circle were only very rarely mentioned as having played the role of social sharing partners. Professionals (e.g., priests, physicians, teachers, psychologists), unfamiliar, or unknown persons are not likely to be selected for this role. However, within the circle of eligible intimates, gender and age differences in chosen partner were observed.

Among adolescents, family members (predominantly parents) were by far the most frequently mentioned partners for the social sharing of emotion for both men and women. In this age group, the second most frequently mentioned partners were friends, who collected about one third of the emotional confidences. The category "spouse or partner" was only rarely mentioned among adolescents, either because they had no partner yet, or because such persons were not yet seen as eligible social sharing partners.

The social sharing network of young adults was markedly different from the adolescents' one. First, the importance of the role played by family members was dramatically lower. This was especially true among male subjects. The percentage of subjects who mentioned family members as social sharing partners dropped from 63.2 at adolescence to 21.8 at young adulthood. Second, for subjects of both genders, spouses and partners emerged as important actors on the social sharing stage. Finally, friends were mentioned about as often as was the case in adolescents data. Thus, this age group appears to be characterized by a quite diversified social sharing network, with three types of partners of about equal importance.

Among older adults, the role played by family members in the social sharing of emotion was again markedly lower. This may be partially due to the fact that parents are no longer available. Additionally, a considerable drop in the importance of friends was registered for male subjects, but not for female ones. However, the most striking fact in this age group was the predominance of spouses and partners in the social sharing processes. This was especially true among men who, at this age, shared their emotions with a spouse or partner in more than 75% of the cases. Data collected on elderly people simply replicated this pattern.

It appears that, although males and females have a very similar social sharing network at adolescence, these networks evolve quite differently. Among females, a differentiated network develops and is maintained up to older adulthood, in which the spouse or partner plays a privileged role. This is not at all the case among males, whose social sharing network virtually disintegrates at mature adulthood, leaving the spouse or partner as the exclusive sharing partner.

Altogether, the data strongly suggest that important aspects of this social behavior elicited by emotion relate to the attachment process. As was shown by classic work conducted by Harlow (1959) and by Bowlby (1969, 1973), exposure to an emotional situation elicits attachment behavior in youngsters at an early age. Our research revealed that when children have reached the elementary school age, their attachment figures—father and mother—become their natural targets for sharing an emotional experience. Later in age, the circle of sharing targets extends to other intimates—siblings, best friends, spouses or partners—who very likely are substitutes to the early attachment figures. With older age, spouses or partners, who are obvious substitutes for attachment figures, are most often attributed the role of exclusive sharing targets.

Emotional Impact of Exposure to the Sharing of Emotion

There is a good deal of empirical evidence suggesting that exposure to the narrative of the emotional experience of another person may induce considerable emotional changes in the listener (Archer & Berg, 1978; Lazarus, Opton, Monikos, & Rankin, 1965; Shortt & Pennebaker, 1992; Strack & Coyne, 1983). Consistent with the view that emotional exposure elicits social sharing behaviors, it can thus be predicted that once exposed to the social sharing of an emotion, receivers will in turn socially share the episode with a third person. In other words, a process of "secondary social sharing," as we called it, is expected to be developed by the listener once he/she leaves the situation in which the primary social sharing process occurred.

Such a prediction sounds counterintuitive, as socially sharing an emotional experience is generally viewed as an intimate matter that assumes confidentiality. The assumption is generally strengthened by the fact that the addressee of the social sharing is an intimate person. Thus, common sense would assume that "secondary social sharing" would rather be the exception than the rule. Yet, the rationale followed here leads to predict the contrary. Once an emotion is shared, secrecy may well be the exception rather than the rule.

This hypothesis was tested by Christophe & Rimé (1997) using a procedure based on the "autobiographic" method. Participants had to retrieve a recent episode in which someone socially shared with them an emotional episode resembling those on a list of 20 events. Three different versions of this list were prepared for the study. One was composed of 20 events that ranked at the lower end of the list of life events developed by Holmes and Rahe (1967) (e.g., quarrel with a friend; death of a pet; unhappy love affair) and was intended to form a list of low emotional intensity events. A second list comprised 20 events that ranked at the higher end of the Holmes and Rahe list (e.g., abortion; divorce; academic or professional failure) and represented a list of moderate emotional intensity events. A third list was composed of 20 events taken out of those mentioned by Green (1990) as potential elicitors of post traumatic stress disorder (e.g., sudden death of a close one; rape; exposure to disaster) and thus constituted a list of high intensity emotional events. A between-subject design was created by randomly assigning each participant to one of these lists. Once participants had retrieved the required episode, they first briefly described it. They then completed a questionnaire covering the dependent measures. Dependent variables included information about the person who was the source of the social sharing, assessment of felt emotions and responses manifested while exposed to the social sharing, and secondary social sharing behavior (i.e., "did the receiver in turn talk about the socially shared episode with a third person later?"). A total of 121 persons, 23 males and 98 females aged between 18 and 55 years ($M = 21.1$), completed the forms of the study.

As regards elicited emotions, the data confirmed that listening to another's emotional story is in itself an emotion event. Indeed, people reported having felt intense emotions while listening to the narrative of the emotional episode. On a 10-point scale assessing emotional impact, low intensity episodes were rated as less impactful ($M = 6.29$) than moderate ($M = 7.83$) or high ($M = 8.08$) intensity ones. The last two ratings were not significantly different. A remarkable feature characterized results from the assessment of felt emotions while listening to social sharing. In all three conditions, subjects rated the emotion of "interest" at nearly maximal level. This suggests that people are very willing to act as listeners of emotional episodes that occur to others.

Further examination of the profile of basic emotions revealed notable effects of the emotional intensity conditions. The more intense the episode, the more it elicited negative emotions like fear, sadness, or disgust. It may be that with more intense episodes, the general openness to listen, as evidenced from ratings of interest, would be counteracted by the raising up of strong negative feelings.

In this sense, the receiver of the social sharing may be trapped in a kind of approach–avoidance conflict. With very intense episodes, avoidance is likely to predominate, with the potential consequence that people may attempt to elude others' confidences. However, when the addressee is an attachment figure or a substitute of such a figure, commitment may be such that listening will generally be granted.

Listeners' Responses

Participants also completed a questionnaire on their reactions while exposed to the social sharing of an emotional episode. The "listener's responses" questionnaire assessed five dimensions, including social support, nonverbal comforting, concrete actions, dedramatizing, and verbalizing. Three of these dimensions significantly discriminated the high emotional condition from the other two. As compared to participants in the low and in the moderate intensity condition, those in the high emotional intensity condition reported less verbalizing and dedramatizing, and more nonverbal comforting. In other words, when intense emotions are shared, listeners reduced their use of verbal mediators in their responses. As a substitute, they manifested nonverbal comforting behaviors, like hugging, kissing, or touching. This suggests that the sharing of an intense emotional experience can decrease the physical distance between two persons.

In summary, the social sharing of emotion has the power of bringing the sender and the receiver closer to one another. As this process develops most of the time with intimates, it can be concluded that it may be instrumental in refreshing and maintaining important social bonds. When circumstances are such that social sharing develops with non-intimates, then it can be expected that a higher degree of intimacy will develop between the interactants.

Secondary Social Sharing

Do receivers share the episode with a third party? Results showed that secondary social sharing occurred in 78.5% of the cases, with no significant difference between conditions of emotional intensity. However, consistent with the observed relationship between the emotional intensity of an episode and the extent of its social sharing, significant differences appeared as a function of emotional intensity of the shared episode. Respondents who recalled a high emotional intensity sharing reported secondary social sharing more often than those who recalled a moderate or a low intensity sharing, these latter two groups failing to differ significantly. Thus, more intense emotional episodes elicited more frequent repetitions of secondary social sharing, and with more new partners. It is remarkable that across conditions, secondary social sharing occurred "at least three or four times" in 37.2% of the cases, and more than six times in 13.2% of the cases. In the high emotional intensity episode, the corresponding figures were even higher, respectively 57.9% and 21.0%. Regarding the number of different partners receiving the secondary social sharing, 35.5% of all subjects chose the item "at least three to four persons", as did 47.4% of those in the high emotional intensity condition.

Thus, these results support the view that the social sharing of emotion leads to the spreading of emotional information. The high figures recorded for secondary social sharing in this study are striking because, in everyday life, we consider our emotional experiences as personal matters. When we socially share such an experience with intimates, we don't usually expect them to repeat it to other people. Yet, they do. And the more emotionally intense the experience, the more likely they are to share it with others.

We stressed that people—especially intimates—are not expected to tell others about emotional events confided to them. The reviewed studies show that this implicit social norm is widely transgressed. At the very least people would be expected to preserve the anonymity of the original source in the process of secondary social sharing. Christophe and Di Giacomo (1995), however, found evidence to the contrary. Respondents in their study reported having revealed the identity of the source in 73% of the investigated cases. Petronio and Bantz (1991) found that people are aware of the receivers' urge to tell confidential information to others. Commonly, disclosers therefore use a prior restraint phrase such as "Don't tell anybody..." in order to minimize the ramifications of disclosure of private information. Nevertheless, the study showed that when information was highly or moderately private, receivers were more likely to tell it than disclosers expected, regardless of the restraint phrase.

To conclude, the collected data confirmed our paradoxical hypothesis that, because social sharing elicits emotions in the listener, listeners themselves engage in secondary social sharing. It seems wise to recommend that if one does not want one's emotional experience to be spread around, one should better not share it at all.

Propagation of Shared Emotions

The data just reviewed showed that once an emotion is shared, there is a high probability that the target will share this emotion with other persons. Now, this opens the possibility that this process will extend even further. Indeed, a target of the secondary social sharing may also experience some emotion when listening, which will elicit in this person the need to share the episode further, leading to a "tertiary social sharing" process. This phenomenon was clearly evidenced by Christophe (1997). In fact, emotions heard in a secondary sharing are shared again with one or several persons in most cases.

Altogether, the data revealed that emotional episodes propagate very easily throughout social networks. It can be calculated that when some intense emotional experience affects a given individual in a community, 50 to 60 members of this community will be informed of the event within the next hours by virtue of the propagation process. From intimates to intimates, and despite any recommendation not to repeat the information, most people in the community will know what happened to one of them. This process of propagation of emotional information has important potential implications in many respects. In particular, it means that emotion elicits intra-group communication. It means that members of a community keep track of the emotional experiences that affect the life of their peers. It also means that in a group, the shared social knowledge about emotional events and emotional reactions is continuously updated as a function of new individual experiences.

The notion that emotional experiences feed collective knowledge of emotion was examined by Harber (K. D. Harber, personal communication, December 1995) under the concept of "the human broadcaster." He proposed that the compulsion to communicate emotional experiences serves not only the teller's needs (for meaning, perspective, closure, etc.) but also the community's need for news. Thus, the teller's urge to disclose supplies the receiver with useful knowledge. One of the predictions of the human broadcaster model is that a major event will "travel" farther than a minor event. A story travels when listeners are so disturbed that they, in turn, find disclosure outlets who may also find it emotionally necessary to tell others. A field study conducted by Harber and Cohen (K. D. Harber, personal communication, December 1995) provided support for these predictions. Students visited a hospital morgue during a psychology class field trip. Later, these students were asked to contact people to whom they relayed their morgue experience and to find out how many people these primary contacts had themselves contacted. Students additionally contacted one of their contact's contacts, and found out how many times these people disclosed. Results showed that a story's travel is positively related to the student's amount of trip ruminations, trip communications, vividness of disclosure, and degree of emotional upset during the trip. These factors correlated not only with how many people the student told the story to, but also with the number of people the student's contact's contacts told it to. All in all, the results from this study—conducted in naturalistic conditions—were remarkably consistent with those of the investigations described earlier.

CONCLUSION: THE PARADOX OF SHARING EMOTION

Research reviewed in this chapter reveals that emotion has marked social and communicative consequences. Indeed, people share their emotional experiences with other persons quite generally. We would like to conclude this chapter by discussing the motives that underlie this behavior. We know from contemporary theories of emotion that when the memory of an emotional episode is re-accessed, the various components of the corresponding emotional reaction (i.e., physiological, sensorial, experiential) are reactivated too (e.g., Bower, 1981, Lang, 1983; Leventhal, 1984).

That this happens in the course of a sharing situation was documented by Rimé, Noël, et al. (1991). They conducted a study in which 96 students had first to describe in detail a past emotional experience (of joy, anger, fear, or sadness, according to the condition) and then answered items exploring what they had experienced while sharing. Nearly all the participants reported to have experienced mental images of the emotional event during the questionnaire-induced social sharing of an emotion. Reports of feelings and bodily sensations were only slightly less frequent. Thus, an overwhelming majority of participants experienced all three elements during the questionnaire-induced social sharing of an emotion.

Participants' affective reaction to the study-induced emotional evocation was assessed by a further question: "To what extent completing this questionnaire was a pleasant or painful experience for you?" The type of primary emotion involved markedly influenced the answers. Not surprisingly, reporting an experience of joy was rated as more pleasant than reporting an emotion of sadness, fear, or anger. However, more surprising was the fact that reporting fear, sadness, or anger was rated as painful or extremely painful only by a minority of the participants. Notwithstanding the reactivation of vivid images, feelings, and bodily sensations of a negative emotional experience, the sharing did not appear as aversive as one could have expected.

Before they would leave, a final question was addressed to the participants of this study: "If you were asked now, would you be ready to complete a similar questionnaire on another past emotional event of the same kind: yes or no?" The answers were observed to be exactly the same whether the emotion shared previously was positively valenced (joy) or negatively valenced (fear, sadness, or anger). Overall 93.7% of the participants gave a positive answer.

These data suggest the paradoxical character of social sharing situations. On the one hand, social sharing reactivates the various components of the emotion that, in the case of negative emotion, should be experienced as aversive. On the other hand, sharing an emotion, whether positive or negative, is a situation to which people are inclined quite willingly. If people are so willing to engage in a social sharing process despite the reactivation of negative feelings, the question arises of what the determinants of social sharing are.

Emotion occurs when events are appraised as discordant with goal orientations, expectations, assumptions, beliefs, or views of the self and the world in general (Janoff-Bulman, 1985; Marris, 1958; Parkes, 1972; Tait & Silver, 1989). Social sharing would allow people to work through the emotional experience, facilitating the restoration of beliefs as well as the search for acceptable meaning to the event (e.g., Silver, Boon, & Stones, 1983; Silver & Wortman, 1980; Tait & Silver, 1989). Cognitive dissonance research showed that people who engage in dissonance reduction typically initiate communication (e.g., Festinger, Riecken, & Schachter, 1956). Social sharing can be hypothesized to play some role in completing the cognitive business elicited by the emotion and thus be well predicted by the need to find a meaning elicited by an event.

In support of this interpretation, our comparison of shared and secret memories (Finkenauer & Rimé, 1998) revealed that secret memories elicited globally more cognitive effort than shared events. Indeed, as compared to shared emotional memories, non-shared memories were associated with (a) greater search for meaning, (b) greater efforts at understanding what had happened, and (c) greater attempts at "putting order in what happened." Items of this kind were later included in various studies in which the memory of an emotional experience was investigated some time after the occurrence of the event. In each of these studies, participants also rated if they still had the need to talk about this memory. In every case, the data supported the view that sharing is related to the need to complete emotion-related cognitive business. This "need for completion" or "search for meaning" appears in the present state of the research as the best predictor of social sharing.

REFERENCES

Archer, R. L., & Berg, J. H. (1978). Disclosure reciprocity and its limits: A reactance analysis. *Journal of Experimental Social Psychology, 14*, 527-540.

Banham, K. M. (1951). Senescence and the emotions: A genetic theory. *Journal of Genetic Psychology, 78*, 183.

Boca, S., Rimé, B., & Arcuri, L. (1992, January). *Uno studio longitudinale di eventi emotivamente traumatici* [A longitudinal study of traumatic emotional events]. Paper presented at the Incontro Annuale delle Emozioni, Padova, Italy.

Bower, G.H. (1981). Mood and memory. *American Psychologist, 36*, 129-148.

Bowlby, J. (1969). *Attachment and loss: Vol. 1. Attachment.* New York: Basic Books.

Bowlby, J. (1973). *Attachment and loss: Vol. 2. Separation: Anxiety and anger.* New York: Basic Books.

Carstensen, L. L. (1987). Age-related changes in social activity. In L. L. Carstensen & B. A. Edelstein (Eds.), *Handbook of clinical gerontology* (pp. 222-237). New York: Pergamon Press.

Carstensen, L. L. (1991). Socioemotional selectivity theory: Social activity in life-span context. In K. W. Schaie (Ed.), *Annual review of gerontology and geriatrics* (Vol. 11, pp. 195-217). New York: Springer.

Carstensen L. L. (1993). Motivation for social contact across life span: A theory of socioemotional selectivity. In J. Jacobs (Ed.), *Nebraska symposium on motivation: Vol. 40. Developmental perspectives on motivation* (pp. 209-254). Lincoln: University of Nebraska Press.

Christophe, V. (1997). *Le partage social des émotions du point de vue de l'auditeur.* [Social sharing of emotion on the side of the target.] Unpublished doctoral dissertation, Université de Lille III, France.

Christophe, V., & Di Giacomo, J. P. (1995). *Contenu du partage social secondaire suite à un épisode émotionnel négatif ou positif* [The content of secondary social sharing following negative and positive emotional episodes.] Unpublished manuscript, Université de Lille III, France.

Christophe, V., & Rimé, B. (1997). Exposure to the social sharing of emotion: Emotional impact, listener responses and secondary social sharing. *European Journal of Social Psychology, 27,* 37-54.

Cumming, E., & Henry, W. E. (1961). *Growing old: The process of disengagement.* New York: Basic Books.

Digman, J. M. (1990). Personality structure: Emergence of the five-factor model. In M. R. Rosenzweig & L. W. Porter (Eds.), *Annual review of psychology* (Vol. 41, pp. 417-440). Palo Alto, CA: Annual Reviews.

Festinger, L., Riecken, H. W., & Schachter, S. (1956). *When prophecy fails.* Minneapolis: University of Minnesota.

Finkenauer, C., & Rimé, B. (1998). Socially shared emotional experiences vs. emotional experiences kept secret: Differential characteristics and consequences. *Journal of Social and Clinical Psychology, 17,* 295-318.

Goldberg, L. R. (1990). An alternative "description of personality": The Big Five structure. *Journal of Personality and Social Psychology, 59,* 1216-1229.

Green, B. L. (1990). Defining trauma: Terminology and generic stressor dimensions. *Journal of Applied Social Psychology, 20,* 1632-1642.

Harlow, H. F. (1959, June). Love in infant monkeys. *Scientific American, 200,* 68-74.

Holmes, T. S., & Rahe, R. H. (1967). The social readjustment scale. *Journal of Psychosomatic Research, 11,* 103-111.

Janoff-Bulman, R. (1985). The aftermath of victimization: Rebuilding shattered assumptions. In C. R. Figley (Ed.), *Trauma and its wake* (pp. 15-35). New York: Brunner/Mazel.

Labouvie-Vief, G., & Blanchard, F. (1982). Cognitive aging and psychological growth. *Aging and Society, 2,* 183-209.

Labouvie-Vief, G., & Devoe, M. R. (1991). Emotional regulation in adulthood and later life: A developmental view. In K. W. Schaie (Ed.), *Annual Review of Gerontology and Geriatrics* (Vol. 11, pp. 172-194). New York: Springer.

Lang, P. J. (1983). Cognition in emotion: Concept and action. In C. Izard, J. Kagan, & R. Zajonc (Eds.), *Emotion, cognition, and behavior* (pp. 192-226). New York: Cambridge University Press.

Lazarus, R. S., Opton, E. M., Monikos, M. S., & Rankin, N. O. (1965). The principle of short circuiting of threat: Further evidence. *Journal of Personality, 33,* 307-316.

Leventhal, H. (1984). A perceptual-motor theory of emotion. In L. Berkowitz (Ed.), *Advances in experimental social psychology* (Vol. 17, pp. 117-182). New York: Academic Press.

Luminet, O., Bouts, P., Delie, F., Manstead, A. S. R., & Rimé, B. (2000). Social sharing of emotion following exposure to a negatively valenced situation. *Cognition and Emotion, 14,* 661-688.

Luminet, O., Zech, E., Rimé, B., Wagner, H. (2000). Predicting cognitive and social consequences of emotional episodes: The contribution of emotional intensity, the five factor model, and alexithymia. *Journal of Research in Personality, 34,* 471-497.

Marris, P. (1958). *Widows and their families.* London: Routledge & Kegan Paul.

Mesquita, B. (1993). *Cultural variations in emotions.* Unpublished doctoral dissertation, University of Amsterdam, the Netherlands.

Parkes, C. M. (1972). *Bereavement: Studies of grief in adult life.* London: Tavistock.

Petronio, S., & Bantz, C. (1991). Controlling the ramifications of disclosure: "Don't tell anybody but..." *Journal of Language and Social Psychology, 10,* 263-269.

Rimé, B. (1989). Le partage social des émotions. [The social sharing of emotions.] In B. Rimé, & K. R. Scherer, (Eds.), *Les émotions* (pp. 271-303). Neufchâtel and Paris: Delachaux et Niestlé.

Rimé, B., Dozier, S., Vandenplas, C., & Declercq, M. (1996). Social sharing of emotion in children. In N. Frijda (Ed.), *ISRE 96. Proceedings of the IXth conference of the International Society for Research in Emotion* (pp. 161-163). Toronto: ISRE.

Rimé, B., Finkenauer, C., Luminet, O., Zech, E., & Philippot, P. (1998). Social sharing of emotion: New evidence and new questions. In W. Stroebe and M. Hewstone (Eds.), *European review of social psychology* (Vol. 9, pp. 145-189). Chichester, UK: Wiley

Rimé, B., Finkenauer, C., & Sevrin, F. (1995). *Les émotions dans la vie quotidienne des personnes âgées: Impact, gestion, mémorisation, et réevocation* [Emotions in everyday life of the elderly: Impact, coping, memory and reactivation.] Unpublished manuscript, University of Louvain, Louvain-la-Neuve, Belgium.

Rimé, B., Mesquita, B., Philippot, P., & Boca, S. (1991). Beyond the emotional event: Six studies on the social sharing of emotion. *Cognition and Emotion, 5,* 435-65.

Rimé, B., Noël, M. P., & Philippot, P. (1991). Episodes émotionnels, réminiscences mentales et réminiscences sociales [Emotional episodes, mental remembrances and social remembrances.] *Cahiers Internationaux de Psychologie Sociale, 11,* 93-104.

Rimé, B., Philippot, P., Boca, S., & Mesquita, B. (1992). Long-lasting cognitive and social consequences of emotion: Social sharing and rumination. In W. Stroebe & M. Hewstone (Eds.), *European review of social psychology* (Vol. 3, pp. 225-258). Chichester, UK: Wiley.

Rimé, B., Philippot, P., Finkenauer, C., Legast, S., Moorkens, P., & Tornqvist, J. (1994). *Mental rumination and social sharing in current life emotion.* Unpublished manuscript, University of Louvain at Louvain-la-Neuve, Belgium.

Rimé, B. , Yogo, M., & Pennebaker, J. W. (1996). *Social sharing of emotion across cultures.* Unpublished raw data.

Shortt, J. W., & Pennebaker, J. W. (1992). Talking versus hearing about the Holocaust experiences. *Basic and Applied Social Psychology, 1,* 165-170.

Sifneos, P. E. (1973). The prevalence of alexithymic characteristics in psychosomatic patients. *Psychotherapy and Psychosomatics, 22,* 255-262.

Silver, R. L., Boon, C., & Stones, M. H. (1983). Searching for meaning in misfortune: Making sense of incest. *Journal of Social Issues, 39,* 81-102.

Silver, R. L., & Wortman, C. B. (1980). Coping with undesirable life events. In J. Garber & M. E. P. Seligman (Eds.), *Human helplessness: Theory and applications* (pp. 279-340). New York: Academic Press.

Singh-Manoux, A., & Finkenauer, C. (in press). Cultural variations in social sharing of emotions: An intercultural perspective on a universal phenomenon. *Journal of Cross-cultural Psychology.*

Strack, S., & Coyne, J. C. (1983). Shared and private reaction to depression. *Journal of Personality and Social Psychology, 44,* 798-806.

Tait, R., & Silver, R. C. (1989). Coming to terms with major negative life events. In J. S. Uleman & J. A. Bargh (Eds.), *Unintended thought* (pp. 351-381). New York: Guilford.

Tangney, J. P. (1991). Moral affect: The good, the bad, and the ugly. *Journal of Personality and Social Psychology, 61,* 598-607.

Taylor, G. J., Bagby, R. M., & Parker, J. D. A. (1997). *Disorders of affect regulation. Alexithymia in Medical and Psychiatric Illnesses.* Cambridge, MA: Cambridge University Press.

Vergara, A. (1993). *Sexo e identidad de genero: Diferencias en el conomiento social de las emociones en el modo de compartirlas.* [Sex and gender identity: Differences in social knowledge on emotions and in ways of sharing them.] Unpublished doctoral dissertation, Universidad del Pais Vasco, San Sebastian, Spain.

Yogo, M., & Onoe, K. (1998, August). *The social sharing of emotion among Japanese students.* Poster session presented at ISRE '98, the Biannual conference of the International Society for Research on Emotion. Wuerzburg, Germany.

– 10 –

The Language of Fear: The Communication of Intergroup Attitudes in Conversations About HIV and AIDS

Jeffrey Pittam and Cynthia Gallois
The University of Queensland

AIDS, the transmission of HIV, and the whole issue of safe sex are above all emotionally-arousing topics. The nature and history of the AIDS pandemic, particularly in Western countries, have meant that AIDS has never been considered simply as a serious disease. Rather, its life-threatening nature, the lack of a cure and the consequent importance of prevention, and its association in the West with minority and stigmatized groups have produced a construction of this disease as a moral issue with a focus of social identity. The mass media, whether in public service announcements (PSAs) on prevention or in popular discussion, have contributed to this construction of AIDS and HIV in mainly emotional terms. When one considers this context, it is surprising that there has been so little discussion of the emotional language around this issue.

In this chapter, our focus is on the emotional language occurring in discussions about AIDS, HIV, and safe sex. Such discussions have from the beginning been framed in terms of social identity and intergroup relations, and HIV transmission has become a prime site for the expression of prejudice (positive as well as negative) based on sexual orientation, ethnic origin, or gender (e.g., Pittam & Gallois, 1996, 1997, 2000). For this reason, we have taken an explicitly intergroup perspective, and examined the emotional language in discourse about AIDS/HIV and safe sex by young heterosexual adults, as a function of their group memberships. In particular, we have looked at the specific emotion words used by these people and the way these choices reflect normative behavior in their salient social groups. These discussions were about AIDS and HIV, and did not necessarily involve the expression of felt emotions. Thus, the chapter concerns language *about* rather than *of* emotions. After a review of key literature and theory, we present the results of a content and discourse analysis focusing on emotional language generically related to fear, which is the dominant emotion in the construction of AIDS in the media and in popular discourse.

The literature on emotion is huge, and has been the subject of comprehensive theoretical reviews (e.g., Frijda, 1986; Scherer, 1988) and of critical reviews of specific areas, including the relation of cognition and emotion (for a recent example, see Forgas & Vargas, 2001). Emotional language has attracted its share of the coverage (for overviews, see Caffi & Janney, 1994; Russell, Fernández-Dols, Manstead, & Wellenkamp, 1995). As it is impossible to do justice to this literature in a single chapter, we mention only the themes that are most relevant to our research: the status of emotion words, the experience of emotion versus discourse about emotion, and the intergroup versus interpersonal nature of emotional language.

STATUS OF EMOTION WORDS

The issue of whether there is a set of culturally universal basic emotions, from which all other emotions are derived, is as old as the topic itself. Darwin (1872/1965), in his seminal work on facial expressions of emotion, appears to have taken the view that a basic set of emotions are associated with specific facial movements in humans and other primates, and that this association has been selected for throughout evolutionary history. Darwin's views have been a strong influence on the debate about the status of emotions, particularly when the experience and expression of emotions by individuals has been considered.

A considerable part of this work questions whether emotions are best described by a small basic set of specific labels or by a small number of prototypical dimensions (e.g., for reviews see Gallois, 1993, 1994; Kövecses, 1995; Oatley & Johnson-Laird, 1996; Russell, 1991). Central to this debate is the issue of the universality of emotion words: whether such words have the same meaning in all languages (e.g., Ortony & Turner, 1990; Russell, 1991) and whether the expression of an emotion is universal across all cultures (e.g., Ekman, 1992). At this stage, there is still no clear resolution to this debate, in spite of extensive research. What is clear is that the scope of emotion words, like the scope of most words, is different in different languages, and that the dimensions originally captured in the semantic differential (Osgood, Suci, & Tannenbaum, 1957) are cross-culturally valid, at least for forced-choice questionnaires. It is also clear that the position a researcher takes in this debate is strongly linked to the methodology employed in research and to the specific research questions, which may preclude a resolution forever. For the purposes of this chapter, it is enough to note that people in all cultures have words for specific emotions and tend to agree within a culture on their denotation and connotation (see Russell et al., 1995).

EXPERIENCE OF EMOTION OR DISCOURSE ABOUT EMOTION

Darwin's influence has also been strong when research on the experience of emotion is considered. For example, most theories of emotional experience assume that emotions and their expression are adaptive in an evolutionary sense, helping animals to address major stressors and novel elements in their environment. Much but not all of this work distinguishes various phases in the appraisal and labeling of emotion (cf. Ekman, 1982: Frijda, 1986; Scherer, 1988), and assumes that the experience of emotion involves a constellation of bodily states, feelings, behaviors, and language labels. Not surprisingly, given its evolutionary focus, most of this research explores the expression of emotion through nonverbal behavior (especially facial expression and paralanguage), reserving the examination of language to the labels associated with each emotion.

There is very strong evidence for cross-cultural similarity or universality in the experience of emotion (Mesquita & Frijda, 1992). In retrospective reports, the same bodily states are associated with, for example, fear, happiness, sadness, anger, and disgust, across a large number of cultures (e.g., Scherer, 1988). In addition, there is a high level of recognition of facial and vocal expressions of emotions across cultures, although once again these results may be dependent on the methodology used (cf. Parkinson, 1996; Wierzbicka, 1995) . There is also evidence that the recognition of emotional expression within a single culture involves social skills that differ between the sexes (Brody, 1996; Noller, 1993) and across individuals (Burgoon, Stern, & Dillman, 1995).

A number of scholars have begun to question the assumption that emotional experience and expression are always adaptive. In fact, there is evidence that in everyday language, emotions are not always described in positive terms. For example, Parrott (1995) found that most connotations of being "emotional" were negative for his participants; being emotional tended to mean losing control and feeling anger or other negative emotions. In this same vein, Oatley and LaRocque (1995) found that being emotional tended to follow a significant breakdown in planning, again suggesting that the emotions people label as such involve negative states. In her study of U.S. women, Lutz (1990, 1996) found that her participants believed that they could not control their own emotions as well as men could, and that they were therefore more emotional than their husbands and male friends. Finally, Crawford, Kippax, Onyx, Gault, and Benton (1992), in a detailed study of reports and interpretations of emotional experience in childhood, found that the women in their study remembered emotions in more negative terms than did men. None of this suggests that people discuss emotions in everyday contexts as though they were helpful in adapting to the environment.

Of course, these results may show no more than the well-known tendency to search for the causes of negative rather than positive events (cf. Heider, 1958), but they do indicate that much of the research on emotional experience may not tap very well into discourse *about* emotion. Thus, it is worth considering the role of emotional language in everyday discourse. What words do people use when they describe emotions they have felt? How do they distinguish in language between emotions, and among the shades of a single emotion? What differences are there between the language used to express emotions and to talk about them? Rather than entering into the debate about the evolution of emotions and emotional language, in this chapter we explore the linguistic choices made by young Australians around one generic emotion, fear.

EMOTION AS INTERPERSONAL OR INTERGROUP

Virtually all studies of emotional communication have considered emotions as individual or interpersonal phenomena (see Gallois, 1993, 1994; Planalp, 1999). This is understandable, given that the experience of emotion is an intrapersonal phenomenon, and the communication of emotion usually takes place in the context of an interpersonal relationship, often dyadic and close. Even studies of the impact of emotion or mood on social judgments have usually involved judgments about the self or about general social issues. A partial exception appears in the work of Forgas and his colleagues (e.g., Forgas, 1995; Forgas & Vargas, in press), who have shown that a negative mood leads to more negative judgments of couples of mixed ethnicity, as well as to more negative political and other social judgments. Nevertheless, it is also important to examine the intergroup aspects of emotional language and expression, because emotion is a key constituent of social identity and thus of intergroup relations.

In his original version of social identity theory (SIT), Tajfel (1982) laid out an essentially affective description of the impact of group memberships on their members. He began from personal identity, which is focused on the need of individuals to compare themselves to salient others in order to achieve an understanding of themselves as distinctive in positive ways (Festinger, 1954). Tajfel argued that the same need to understand the world through the use of categories and to perceive oneself in positive and distinctive terms extended to social identity, or the identity derived from important group memberships. He argued that any characteristic, however arbitrary, may be used to distinguish between an ingroup and an outgroup, and demonstrated that groups could even be formed around arbitrarily assigned characteristics (Tajfel, Flament, Billig, & Bundy, 1971). Naturally occurring group memberships like race, gender, age, and so forth are especially likely to form the basis for important social identities. According to this theory, group members feel a strong loyalty to their ingroups, and strive to compare them to salient outgroups in positive ways.

Later versions of SIT (e.g., Hogg & Abrams, 1988; McGarty, 1999; Tajfel & Turner, 1979; Turner, 1987) have kept to this basic conceptualization, while exploring the contextual conditions that are likely to activate social rather than personal identity. This research has clearly shown that contexts of competition, rivalry, or threat tend to activate social identity, whereas contexts of cooperation and intimacy activate personal and relationship identities. By its very nature, AIDS and HIV transmission produce a context where social identity is activated. First, HIV has been associated with minority groups, so that comparisons like homosexual/heterosexual, drug user/non-user, and the like, spring readily to mind. Secondly, the centrality of sexual intercourse to the transmission of HIV means that gender becomes an immediately salient group membership for heterosexuals, who are potentially threatened by members of the other sex.

In conditions of social identity salience, people try to maximize differences with the salient outgroup (or outgroups), while minimizing individual differences among members of the ingroup and among members of the outgroup (Turner, 1987). Any characteristic, belief, or behavior invoking a distinction with an outgroup that is positive for the ingroup is likely to become an important group marker or prototype. Group members come under normative pressure to behave in accordance with group prototypes. In addition, when social identity is salient, group members may collaborate to form a consensus about prototypical behaviors and beliefs, especially where a leader or other important figure has not already exemplified them. This means that both ingroup and outgroup behavior is important to the construction of group prototypes and to the maintenance of social identity.

Social identity comparisons involve an affective component. Descriptive stereotypes, which are important elements in the distinction between ingroups and outgroups, have an affective as well as a cognitive aspect, and people look for negative stereotypes to use for rival or threatening outgroups, along with positive stereotypes for their ingroups (cf., McGarty, 1999; Oakes, Haslam, & Turner, 1994). For example, in the context of discussions about HIV, Pittam and Gallois (1996, 2000) found that when unsafe sexual behavior was explained by young heterosexuals, members of outgroups were stereotyped in negative terms (uneducated, obsessed with reproduction, enthralled by institutions like the church), whereas ingroup members were stereotyped in more sympathetic terms (seeking love, influenced by sexual desire). Stereotypes influence the construction of prototypical behaviors for the ingroup and prototypical expectations for outgroups, which reinforce the stereotypes.

Given the affective components in social identity, many group prototypes should involve emotions and their expression. When social identity is salient, group members are likely to look for appropriate emotions to express about outgroups and about their own group. Groups are thus likely to develop norms about what emotions to feel in a particular context and about the appropriate words to use and ways to behave in expressing them. Indeed, in their extension of communication accommodation theory, Giles, Coupland, and Coupland (1991) noted the importance of emotional needs in determining linguistic behavior in intergroup communication. We set out to explore the linguistic choices involved in this process in discussions about HIV.

HIV AND EMOTION IN AUSTRALIA

The first AIDS patient in Australia was identified in 1982, and the community response to HIV, largely driven by the gay community, dates from 1984. The Australian Federal government began funding community groups for HIV prevention and education work from the mid-1980s, and launched the first PSA targeted at a national audience in 1987. This PSA soon came to be known as the Grim Reaper ad, and it is likely to be the most fear-oriented AIDS PSA produced anywhere in the world. The ad shows a ten-pin bowling alley with a series of Grim Reapers, complete with scythes, knocking down men, women, children, and babies arranged in family groups. The narrator, in a sepulchral tone of voice, states that AIDS kills not only homosexuals and drug users, but everyone, and ends with the cryptic message: "always use condoms...always." This was intended to put AIDS on the agenda in Australia, and to shake the faith of the general public in the ability of science and the family to protect them from infection, rather than to change behavior. The ad appears to have achieved this aim, as the Grim Reaper ad is still the best-known AIDS PSA today, even though many younger people have never seen it (Lupton, 1994).

To emphasize the scary message, the Grim Reaper ad was followed by a second PSA, directed at a mass television audience but specifying injecting drug users (IDUs), featuring a game of Russian roulette with AIDS as the fatal bullet and the Grim Reaper in the background. A third PSA in the series showed two people about to have sex and falling onto a bed of needles. After this campaign, national PSAs became progressively less fear-related, first using testimonials of non-IDU heterosexuals infected with HIV, then using humor to normalize condom use, and finally emphasizing the importance of not discriminating against people with HIV ("HIV doesn't discriminate —people do"). Since the mid-1990s, PSAs have been targeted at specific groups, using local rather than national media. All the national campaigns, however, have stressed the fact that anyone can be infected with HIV, and that personal responsibility for prevention is crucial.

Even if this had not already been the case, the Grim Reaper ad and its followers would have ensured that AIDS and HIV would be constructed in Australian discourse mainly through fear. In fact, the PSA tapped very well into popular and media discourse at the time and since (Lupton, 1994). Thus, fear has been the overwhelmingly mentioned emotion in all discussions of HIV and AIDS among young people in this country (e.g., see Timmins, Gallois, Terry, McCamish, & Kashima, 1993). AIDS has the status of a dread risk (Weinstein, 1989), in that people describe it as the disease they most fear, but their worry about it does not predict preventive behavior, and young heterosexuals describe themselves as being at very low risk of infection (Timmins, Gallois, McCamish, Kashima, & Terry, 1994; Timmins, Gallois, McCamish, & Terry 1998; Moore & Rosenthal, 1993). Fear appeals through PSAs may or may not produce compliance, depending in part on the other emotions they trigger. For example, fear accompanied by anger may produce reactance rather than compliance (Dillard, Plotnick, Godbold, Freimuth, & Edgar, 1996), whereas fear associated with interest may result in a more active public response (Roser & Thompson, 1995).

Given this context, it makes sense to concentrate on the discussion and expression of emotions related to fear. In this chapter, we describe the fear-related emotions mentioned in discussions of HIV, AIDS, and safe sex among young heterosexual adults in Australia. The analysis examines the fear-related words mentioned and the interpersonal and intergroup contexts in which they appear. In addition, we looked at the specific themes associated with each word, and considered them in an intergroup perspective. Overall, our aim was to shed light on the way these words are used to reflect norms and prototypes in this intergroup context.

METHOD

Speakers and Recordings

In the study reported here we set out to construct social ingroups for conversations, using strangers rather than friends or family members, as both emphasize personal relationships. The social categories of age, ethnicity, and being a student were homogeneous, but gender of speakers was varied, as it is the major social category linked to situations where outgroup members are potential sexual partners for heterosexuals.

Seventy-two female and 60 male first-year university students (17–24 years) participated, and all reported that they were Australian-born, of Anglo-Celtic ethnicity, were exclusively heterosexual, and did not know the other members of the group in which they participated. They were formed into 33 groups of four people in differing gender combinations. This resulted in 13 female same-gender groups, 10 male same-gender groups, and 10 mixed-gender groups (two males and two females in each). We used groups of four because we were concerned with group talk and wanted participants to have the opportunity to form alliances with other group members in positioning themselves relative to the issues discussed. This strategy also gave the opportunity for gender ingroups and outgroups to be formed in the mixed-gender sessions.

Each group took part in a discussion of HIV, AIDS, and related matters. Conversations were audio recorded, and lasted for as long as the participants wanted. In the event, this ranged from 15 to 40 minutes. Two trained facilitators (one female and one male) were used to initiate the conversation ("What do you think of when you hear the word AIDS?") and supply very generally worded topics when and if necessary. Beyond this, facilitators did not take part in the discussions. Facilitators were of the same ethnic background and age as the participants, and of the same gender in the same-gender groups. Male and female facilitators were used alternately for the mixed-gender groups.

All recordings were transcribed by two researchers working independently. Disagreements were resolved by both transcribers listening to the segment in question and reaching agreement on the words used. When disagreements could not be resolved, the transcription was left blank. Extracts used in this chapter are exact transcriptions; dysfluencies and ungrammaticality remain. When interruptions or simultaneous speech occurred, we have shown the point where it started with a square bracket. An equal sign at the end of a line indicates that the speaker continued straight on with no pause; the duration of pauses are shown in brackets to the nearest one-tenth second. Underlining indicates emphasis; additional comments are shown in double brackets. To preserve anonymity, speakers are referred to by gender and order of speaking in the extract (thus, M1 refers to the first male speaker in the extract).

FINDINGS AND DISCUSSION

We conducted a content analysis on the transcriptions to see how frequently the emotion of fear was raised by our participants. Table 10.1 shows the frequency of the full range of emotion words found in the 33 sessions that are related semantically to fear: a total of 251 instances. To help us further understand the ways in which the words were being used, we checked what (if anything) the participants seemed to fear, who (if anyone) they believed was fearful, and who (if anyone) they believed was feared. These three superordinate categories were then subcategorized, as shown in Table 10.1. It is these subcategories and the three superordinate categories that represent the themes associated with each emotion word. In many instances, one or other of the three superordinate categories was not present. These cases are listed as "unspecified." Often the fear of "catching AIDS" was accompanied by a specified source of the fear (e.g., stepping on a used needle on the beach; the images in PSAs; ineffective

TABLE 10.1

Categorization and Frequency of "Fear" Words

Fear Word	What Feared?						Who Fears?					Who Feared?					Total
	D	TU	A	CA	MF	U	E	I	P	O	U	O_n	O_s	O_w	R	U^a	
Worry[b]	2	3	22	51			19	1	51	7		5	13	6	17	37	78
Scare			2	5	46		2	27	2	10	8	8		12	4	5	34	55
Shock	1		1	7	18	1	2		24		2	1			19	8	28
Scary	5	1		16			3	1	18				3		2	17	22
Fear	1	4		13		2	9		7		4	4	2		1	13	20
Dangerous		1	9			2	5		1	2	4	3	2	1		6	12
Concern			4	5			2	1	5	1					6	3	9
Freaked	1			5	1			1	5	1		1			5	1	7
Afraid			1	5			2		3	1			2		3	1	6
Frightened		2		3			2		2	1			2		2	1	5
Nervous				3			1		2			1	1	1			3
Horrified			1	1			1		1							2	2
Petrified				2					2				2				2
Terrified				2					2						1	1	2
Totals	10	13	43	159	19	7	73	6	133	21	18	15	39	11	62	124	251

[a] D=Death; TU=The Unknown; A=AIDS; CA=Catching AIDS; MF=My Friend; U=Unspecified; E=Everyone; I=Ingroup; P=Personal; O=Outgroup; U=Unspecified; O_n=Outgroup named; O_s=Outgroup strongly implied; O_w=Outgroup weakly implied; R=Relational; U=Unspecified

[b] These represent a generic form of the word; derivatives are included in each form (e.g., freak and freaked) unless a particular derivative makes up a sufficiently large subset that is used somewhat differently to other forms (e.g., scare and scared are included together but scary is listed separately).

condoms). These have not been separately listed in Table 10.1, as such sources of fear did not appear frequently. The aim of Table 10.1 is to provide an indication of the trends of use of fear words by the participants, rather than listing all perceived sources of fear. A distinction has been drawn between "catching AIDS" and "AIDS" itself, although it can be argued, of course, that a fear of catching AIDS underlies both.

Frequently Used Words: "Worry" and "Scare"

As can be seen, the two most frequently used words were "worry" and "scare" and their derivatives. The former was used most often to refer to the fear of catching AIDS but not always fear for oneself: Sometimes the fear was for a close friend or family member. Extract 1, from a female same-gender group, shows the word (highlighted) used in this way. On this occasion the speaker is fearful for her friend.

Extract 1

F1: I'd say **I'd worry a lot** about my friend ((small laugh)) you know once I'm attached to a friend, I like everything to be good for them and everything (1.1) and I think it would probably upset me a lot (2.3) yeah (1.3) probably a lot ((laugh))

Although participants worried *for* a friend in case they became HIV positive, when presented with a hypothetical scenario in which a close friend did have AIDS, they almost exclusively *did not* worry about themselves. Extract 2 is of this kind.

Extract 2

M1: Like if it was like your friend (1.1) you wouldn't **you wouldn't sort of worry** about it 'cause you know that you don't have any chance of catching it

Without wishing to appear cynical about our participants' intended altruism, perhaps a more accurate reaction is shown in the next extract; a response given by a participant in a female same-gender group.

Extract 3

F1: I think I'd stand by them I don't **I don't really think it would worry** me a great deal (1.5) I don't know actually you probably have your first (1.0) reactions of "Oh gosh" you know "I hope I don't catch it" that sort of thing but (1.0) I don't think that would last very long I think that's just an instant reaction that everyone would have

"Scare," on the other hand, although it was used to indicate a fear of catching AIDS, tended to be used more globally, referring to everyone. It was often used when referring to outgroups, although the outgroup named tended to be "people with AIDS." In Extract 4, it is AIDS itself that is feared. The word appears three times in quick succession, although in the first two there is no specified individual or group feared. In Extract 5, it is catching AIDS that is feared, but people are seen as not scared enough.

Extract 4

F1: Yeah I suppose at least everybody knows that *everybody is scared* of AIDS which (0.7) to some extent is good as far as you know people will probably you know use protection when *if they weren't scared of the thing* (0.7) just wouldn't bother (0.6) but then I suppose *they get scared of people who have AIDS* and (0.8) probably shouldn't be

Extract 5

M1: Mm, yeah, the Australian attitude, she'll be right
                                                    [
M2:   *They're just not sc* (0.7) *scared enough*

Importance of Words to Categories

The difference in frequency of use of the 14 words in Table 10.1 distorts the relative importance of each to the categories listed. To account for this we checked the importance of each word to each category. This was calculated as the number of each word in the particular category multiplied by the total number of words in that category, divided by the total number of the word in question multiplied by the total number of all fear words (251). The results are presented in Table 10.2; the two highest values in each category have been underlined.

Using these indices of importance, "worry" is shown to be rather less important to any particular category than its overall frequency of use might suggest, although clearly the overall frequency shows it to be a popular word. It remains of some importance for expressing fear of catching AIDS. Other words had a higher index, including "petrified," "terrified," "nervous," and "scare," but the first three of these appeared in the transcripts on only two or three occasions. "Worry" was also important for the personal category, and particularly when outgroups were weakly implied. Extract 6 is an example in which the weakly implied outgroup is heterosexual men.

TABLE 10.2

Importance of Fear Words to Each Category

Fear Word	What Feared?						Who Fears?					Who Feared?				
	D	TU	A	CA	MF	U	E	I	P	O	U	O$_n$	O$_s$	O$_w$	R	U[a]
Worry[b]	0.64	0.80	1.65	1.03			0.84	0.54	1.23	1.07		1.15	1.07	1.75	0.32	0.94
Scare	5.00	0.76	0.53	1.32		1.31	1.69	1.52	0.35	1.74	2.03		1.41	1.66	0.37	1.23
Shock	5.68		0.21	0.39	10.41	1.55	0.24		1.61		0.99	0.78			3.39	0.58
Scary		0.94		1.14			0.47	1.90	1.53				0.87		0.37	1.53
Fear	1.25	3.13		1.02		3.58	1.54		0.66		2.78	3.57	0.64		0.20	1.29
Dangerous		4.20	1.93			10.21	1.43		0.40	1.98	4.63	10.71	1.07	3.25		1.01
Concern	3.57		2.58	0.87			0.75	4.64	1.06	1.32					2.73	0.66
Freaked				1.12	1.89			5.96	1.36	1.70		2.55			2.34	0.57
Afraid			0.97	1.31			1.13		0.96	1.98			2.13		2.05	0.33
Frightened		8.35		0.94			1.35		0.75	2.38			2.55		1.64	0.40
Nervous				1.57			1.67		1.25				2.14		1.37	
Horrified		2.91	2.91	0.79			1.75		1.00							1.98
Petrified				1.57					1.88				6.41			
Terrified				1.57					1.88						2.05	0.99

[a]D=Death; TU=The Unknown; A=AIDS; CA=Catching AIDS; MF=My Friend; U=Unspecified; E=Everyone; I=Ingroup; P=Personal; O=Outgroup; U=Unspecified; O$_n$=Outgroup named; O$_s$=Outgroup strongly implied; O$_w$=Outgroup weakly implied; R=Relational; U=Unspecified.

[b]These represent a generic form of the word; derivatives are included in each form (e.g., freak and freaked) unless a particular derivative makes up a sufficiently large subset that is used somewhat differently to other forms (e.g., scare and scared are included together but scary is listed separately).

Extract 6

F1: Yeah you've got to also I mean I guess if it's committed y(
have to worry about that that night in the back of ((small laugh
four words)) the Lexington Queen or something ((small laugh over ___t two
words))

Similarly, "scare," although a frequently used word, was important mainly
with respect to fear of catching AIDS, when referring to the fear felt by everyone
and by outgroups, and, like "worry," when weakly implying that outgroups are
to be feared. Extract 7 gives one such example.

Extract 7

M1: You know like a lot of people will wear condoms and (0.3) so they don't
get AIDS but (1.0) like even on this show last night (0.4) um over in Thailand
they were talking to Australians and stuff that were over there and some guy
was saying "Oh" (0.9) they said **"Aren't you scared** of getting AIDS?" and
they were saying "No I can't get AIDS" they just believe they can't get it and
these are guys that are thirty forty years of age

Here the word is used in a negative construction, the outgroup who is not scared
is older (relative to the participant) Australian men and the weakly implied
outgroup is female Thai prostitutes.

Somewhat more interesting, perhaps, is the importance of "scary" when
talking about fear of death:

Extract 8

M1: Yeah it inevitably leads to death **I suppose that's scary**

In general terms we noticed a difference in usage of low intensity emotion
words to that of high intensity words. "Scary" is a less intense word used to
express what might be the ultimate fear. Interestingly, death *per se* appeared
infrequently, and when it did, the low intensity words scary and worry tended to
be used (70% of the time). High intensity words such as "petrified," "terrified,"
"horrified," and "freaked" were used infrequently and almost always to refer to the
fear of personally catching AIDS.

One of the clearest trends was the use of "shock" and its derivatives as a
response to a hypothetical situation in which a close friend became HIV
positive. All but one instance of this scenario invoked a "shock" word (usually
"shocked"), and the word tended to be used most often for this situation. As a
result, it was the most important word for this category and for describing close
relationships. "Shock" is not as closely related to "fear" semantically as the other

words in Table 10.1, and it is included here due to its importance to the situation. Extracts 9 and 10, from different sessions, provide typical examples.

Extract 9

F1: It'd be **such a huge shock** to find out that one of your friends had AIDS or something
        [
F2:   It'd make yeah
           [
F3:   Yeah

Extract 10

F1: **I'd be shocked I'd be shocked** at first but then you know (1.0) be supportive too

Interestingly, along with "shock" as the perceived initial reaction to this particular scenario, "worry" was used to indicate how, in the general situation of knowing that a friend was HIV positive, the speaker would *not* worry. Extracts 2, 3, and 6 above are such examples. Indeed, "worry" was the word most frequently placed in a "not [emotion word]" construction (44 occurrences, 56%), largely due to this general situation. The other word most frequently used in such a construction (scare: 7 instances, 23%) tended to be used to indicate that others weren't scared enough of AIDS (see Extract 5).

Fear and AIDS PSAs

As noted in the introduction, Australian AIDS PSAs have tended to emphasize extreme fear coupled with individual salience and helplessness and, more recently, personalizing people who are HIV positive. The earlier PSAs, and in particular the Grim Reaper campaign, were targeted by our participants as things to invoke fear in themselves. Here, the word "scare" was the most frequently used, often as part of the phrase "scare tactics." The most important words, however, were "shock" and "freak" (the calculations for fear of PSAs do not appear in Table 10.2). Extracts 11 to 13 give examples of each.

Extract 11

F1: Well they're only just coming away from *the scare tactics* really you know the=
                                               [
F2:   Yeah
                                           [
F3:   Mm

F1: =the Grim Reaper (0.3) *scare tactics*

Extract 12

F1: Um (1.0) you don't really see a lot of it (0.5) in like you used to at first when it first came out you'd see a lot of those commercials and they really I mean I really=

          [

F2:   Yeah

             [

F3:   Oh yeah

F1: =when I saw them *they really shocked me*

Extract 13

F1: Yeah and all these people just got (0.3) you know knocked over like skittles and it's like=

             [

F2:   Yeah

F1: =you know I (0.6) *that really freaked me out* that really affected me

Emotional Language and Social Identity

A number of the examples given so far show evidence of intergroupness. Identity, whether referring to an ingroup or outgroup, whether personal or social identity, provides one focus for talking about HIV/AIDS and related matters. As can be seen from Table 10.1, emotional language, when used in the context constructed for this study, may also have this focus. It is noteworthy that the third superordinate category (who is feared?) has more instances categorized as "unspecified" than either of the other two superordinate categories (almost 50%). Despite this, some 65 examples (25.8%) either name or imply an outgroup to be feared. Although we have highlighted the most important words in the three outgroup categories in Table 10.2, no single word can be highlighted as particularly important to outgroups as a whole. That said, of the 65 examples, 29 (44.6%) use low intensity words ("worry," "nervous," and "scary") and 3 (4.6%) use high intensity words ("petrified" and "freaked"). The outgroups tend to be "people with AIDS," but others, such as intravenous drug users and heterosexual men, are also implied. The next two extracts show examples of the last two groups coupled with a high intensity word. In each case it is the possible irresponsible behavior of the groups in question rather than the groups *per se* that invokes the fear word.

Extract 14

F1: *I'm petrified* of that walking along on the beach and (0.4) "Oh no"

Extract 15

F1: Women don't have to worry
                    [
F2:   Not at all they (0.3) doctors always give you a lecture
    now if if you're going on the pill still use a condom I guess
F3: Yeah
F2: It's not (0.4) a condom's not safe
          [
F4:   I would *I'd be petrified*

Usually, however, it is a low intensity word that is used when an outgroup is feared. One unusual example is given in Extract 16.

Extract 16

F1: And the phenomenal amount of pop stars from the seventies who have died from it ((small laugh on last three words)) that *really   gets   m e   nervous*

Why the death of 1970s pop stars should make F1 nervous is never explained. A more typical example is given in Extract 17.

Extract 17

F1: What *I get worried* about when I think about AIDS is um (0.9) is sleaze-buckets ((laugh)) guys who are sleaze-buckets and you think oh you know o h definitely you you=
                                        [
F2:   Yeah
F1: =AIDS carrier you know (0.4) don't wanna (0.5) mess around with that person because of that

Although the meaning is reasonably clear from the extract, F1 shortly explains in greater detail that she labels as sleaze-buckets heterosexual men who are promiscuous.

In terms of who is fearful, we found 21 examples (8.3%) of outgroups named and a few examples of the ingroup (rather than just the self or everyone). Examples of each are given in the following two extracts, respectively.

Extract 18

F1: I don't think that's a big deal because I think *most prostitutes are s o scared* of it anyway that they'll all wear condoms (0.7) 'cause I don't think any man or the money's worth your life

Extract 19

F1: I know a lot of my friends (0.6) I mean that sort of *freaked a lot   o f* (0.4) *us out*

The latter was said once again in the context of a discussion about the Grim Reaper.

GENERAL DISCUSSION

In describing the fear-related emotions used by young Australian participants in discussions about HIV and AIDS, we have seen that only a few words (and their derivatives) appear with any regularity. These include "worry" and "scare/scary" and, to a lesser extent, "shock" and "fear." The most frequent interpersonal context discussed involved the hypothetical case of a close friend who was HIV positive (this was simply because this scenario was raised by facilitators), but a personal fear of "catching AIDS" was the topic most frequently raised by participants.

From an intergroup perspective, there were relatively few instances linked to the superordinate category "Who Fears?". Most of the time, this category was related to an interpersonal perspective. There were far more occasions, however, when outgroups were linked to the third superordinate category: "Who is Feared?". More often than not, the outgroups were implied rather than named.

Remembering that our data involve a particular group of people talking about a particular topic, we can say that fear-related words were the emotion words that appeared most frequently. Although we have not dealt with them here, we checked the transcripts for all emotion words and found that few others appeared more than a very few times. The only partial exception to this was the word "sad," which appeared 20 times across the 33 sessions, nearly always in the context of referring to children who were HIV positive.

The trends toward establishing norms and prototypes for this group and this topic were discussed earlier. The clearest example of a group prototype for expressing a fear-related emotion in this context involves the use of the word "shock" and its derivatives to describe initial reactions to learning that a close friend was HIV positive. This word also occurred in close proximity to statements indicating that speakers would not worry for themselves, even though the greatest fear expressed by the participants in these sessions was that to the self. High and medium intensity words tended to be used when talking about fear for self, but it was the low intensity word "worry" that was used to indicate lack of fear for self. This choice of words, as the extracts show, points to an affective tone of sympathy and concern where personal relationships are involved.

Where outgroups were mentioned or implied, the most important words to the categories were "dangerous" and "frightened," although the word "fear" also had a role here (see Table 10.2). The word "scare" and its derivatives, along with "fear" and "dangerous," was most important in more global contexts where fear of death or AIDS was mentioned. By contrast, "shock" and "concern" (along with the frequently used "worry") were the most important words when interpersonal relationships were discussed. These same words were important when the self or the ingroup were named as the people who were afraid. Thus, participants made a distinction in word choice for personal and interpersonal contexts on the one hand and social identity and global contexts on the other. The words they chose to talk about a general fear of AIDS and death overlapped with the important fear-related words they used to refer to outgroups, but not nearly as much with the words they chose to describe themselves, their ingroup, and their friends. In general, as the extracts also show, their affective tone where outgroups were involved, or where a general or unspecified fear was discussed, was more negative.

In the Australian context, HIV and AIDS have been constructed in terms of personal responsibility and the inappropriateness of blaming others. The participants in this study reflected this message in the reactions they described to the PSAs, as well as in their tendency to talk in generalities and in terms of unspecified people and events. This tendency does not stop blame being laid (see Pittam & Gallois, 1997), but it does mean that in talking about this topic, young heterosexuals in this country link fear to themselves far more than they link it to outgroups. Even so, the results of this study highlight the fact that social identity in terms of gender, nationality, and sexuality, is still important in the choice of words to describe the appropriate emotions toward the self and others when HIV and AIDS are discussed. A more negative attitude towards salient outgroups, thus, may be expressed in the subtleties of particular words rather than by a difference in the emotion expressed. Future research must explicate this relationship further, by taking an intergroup perspective to examine language about fear and other emotions in intergroup contexts.

REFERENCES

Brody, L. R. (1996). Gender, emotional expression, and parent-child boundaries. In R. D. Kavanaugh & B. Zimmerberg (Eds.), *Emotion: Interdisciplinary perspectives* (pp. 139-170). Mahwah, NJ: Lawrence Erlbaum Associates.

Burgoon, J. K., Stern, L. A., & Dillman, L. (1995). *Interpersonal adaptation: Dyadic interaction patterns*. New York: Cambridge University Press.

Caffi, C., & Janney, R.W. (1994). Toward a pragmatics of emotive communication. *Journal of Pragmatics, 22*, 325-373.

Crawford, J., Kippax, S., Onyx, J., Gault, U., & Benton, P. (1992). *Emotion and gender: Constructing meaning from memory.* London: Sage.

Darwin, C. (1965). *The expression of the emotions in man and animals.* Chicago: University of Chicago Press. (Original work published 1872).

Dillard, J. P., Plotnick, C. A., Godbold, L. C., Freimuth, V. S., & Edgar, T. (1996). The multiple affective outcomes of AIDS PSAs: Fear appeals do more than scare people. *Communication Research, 23*, 44-72.

Ekman, P. (Ed.). (1982). *Emotion in the human face* (2nd ed.). Cambridge: Cambridge University Press.

Ekman, P. (1992). Are there basic emotions? *Psychological Review, 99*, 550-553.

Festinger, L. (1954). A theory of social comparison processes. *Human Relations, 7*, 117-140.

Forgas, J. P. (1995). Mood and judgment: The affect infusion model. *Psychological Bulletin, 117*, 39-66.

Forgas, J. P., & Vargas, P. (2001). The role of different processing strategies in mediating mood effects on judgments and cognition. In J. P. Forgas (Ed.), *Feeling and thinking: The role of affect in social cognition.* Cambridge: Cambridge University Press.

Frijda, N. H. (1986). *The emotions.* Cambridge: Cambridge University Press.

Gallois, C. (1993). The language and communication of emotion: Universal, interpersonal, or intergroup? *American Behavioral Scientist, 36*, 262-270.

Gallois, C. (1994). Group membership, social rules, and power: A social-psychological perspective on emotional communication. *Journal of Pragmatics 22*, 301-324.

Giles, H., Coupland, N., & Coupland, J. (Eds.). (1991). *Contexts of accommodation: Developments in applied sociolinguistics.* Cambridge: Cambridge University Press.

Heider, F. (1958). *The psychology of interpersonal relations.* New York: Wiley.

Hogg, M., & Abrams, D. (1988). *Social identifications: A social psychology of intergroup relations and group processes.* London: Routledge.

Kövecses, Z. (1995). Introduction: Language and emotion concepts. In J. A. Russell, J.-M. Fernández-Dols, A. S. R. Manstead, & J.C. Wellenkamp (Eds.), *Everyday conceptions of emotion: An introduction to the psychology, anthropology and linguistics of emotion* (pp. 3-16). London: Kluwer.

Lupton, D. (1994). *Moral threats and dangerous desires: AIDS in the news media.* London: Taylor & Francis.

Lutz, C.A. (1990). Engendered emotion: Gender, power, and the rhetoric of emotional control in American discourse. In C. A. Lutz & L. Abu-Lughod (Eds.), *Language and the politics of emotion* (pp. 69-91). Cambridge and Paris: Cambridge University Press and Edition de la Maison des Sciences de l'Homme.

Lutz, C. A. (1996). Cultural politics by other means: Gender and politics in some American psychologies of emotions. In C. F. Graumann & K. E. Gergen (Eds.), *Historical dimensions of psychological discourse* (pp. 125-144). New York: Cambridge University Press.

McGarty, C. (1999). *Categorization in social psychology.* London: Sage.

Mesquita, B., & Frijda, N. H. (1992). Cultural variations in emotions: A review. *Psychological Bulletin, 112*, 179-204.

Moore, S., & Rosenthal, D. (1993). *Sexuality in adolescence.* London: Routledge.

Noller, P. (1993). Gender and emotional communication in marriage: Different cultures or differential social power? *Journal of Language and Social Psychology, 12*, 132-152.

Oakes, P. J., Haslam, A. S., & Turner, J. C. (1994). *Stereotyping and social reality*. Oxford: Blackwell.

Oatley, K., & Johnson-Laird, P. N. (1996). The communicative theory of emotions: Empirical tests, mental models, and implications for social interaction. In L. L. Martin & A. Tesser (Eds.), *Striving and feeling: Interactions among goals, affect, and self-regulation* (pp. 363-393). Mahwah, NJ: Lawrence Erlbaum Associates.

Oatley, K., & LaRocque, L. (1995). Everyday concepts of emotions following every-other-day errors in joint plans. In J. A. Russell, J.-M. Fernández-Dols, A. S. R. Manstead, & J. C. Wellenkamp (Eds.), *Everyday conceptions of emotion: An introduction to the psychology, anthropology and linguistics of emotion* (pp. 145-164). London: Kluwer.

Ortony, A., & Turner, T. J. (1990). What's basic about basic emotions? *Psychological Review, 97*, 315-331.

Osgood, C. E., Suci, G. J. and Tannenbaum, P. H. (1957). *The measurement of meaning*. Urbana: University of Illinois Press.

Parkinson, B. (1996). Emotions are social. *British Journal of Psychology, 87*, 663-683.

Parrott, W. G. (1995). The heart and the head: Everyday conceptions of being emotional. In J. A. Russell, J.-M. Fernández-Dols, A. S. R. Manstead, & J. C. Wellenkamp (Eds.), *Everyday conceptions of emotion: An introduction to the psychology, anthropology and linguistics of emotion* (pp. 73-84). London: Kluwer.

Pittam, J., & Gallois, C. (1996). The use of narrative in intergroup references in conversations about HIV and AIDS. *Journal of Language and Social Psychology* (special issue on intergroup processes), *15*, 312-334.

Pittam, J., & Gallois, C. (1997). Attribution of blame in conversations about HIV and AIDS. *Communication Monographs, 64*, 201-218.

Pittam, J. & Gallois, C. (2000). Malevolence, stigma, and social distance: Maximizing intergroup differences in HIV/AIDS discourse. *Journal of Applied Communication Research, 28,*, 24-43.

Planalp, S. (1999). *Communicating emotion: Social, moral, and cultural processes*. Cambridge: Cambridge University Press.

Roser, C., & Thompson, M. (1995). Fear appeals and the formation of active publics. *Journal of Communication, 45*, 103-121.

Russell, J. A. (1991). Culture and the categorization of emotions. *Psychological Bulletin, 110*, 426-450.

Russell, J. A., Fernández-Dols, J.-M., Manstead, A. S. R., & Wellenkamp, J. C. (Eds.). (1995). *Everyday conceptions of emotion: An introduction to the psychology, anthropology and linguistics of emotion*. London: Kluwer.

Scherer, K. R. (1988). *Facets of emotion: Recent research*. Hillsdale, NJ: Lawrence Erlbaum Associates.

Tajfel, H. (1982). *Social identity and intergroup relations*. Cambridge: Cambridge University Press.

Tajfel, H., Flament, C., Billig, M. G., & Bundy, R. F. (1971). Social categorisation and intergroup behaviour. *European Journal of Social Psychology, 1*, 149-177.

Tajfel, H., & Turner, J. (1979). An integrative theory of intergroup conflict. In W. G. Austin & S. Worchel (Eds.), *The social psychology of intergroup relations*. Monterey, CA: Brooks/Cole.

Timmins, P., Gallois, C., McCamish, M., Kashima, Y., & Terry, D. (1994). Sources of information about HIV/AIDS, trust of sources, and perceived risk of infection. *Australian Journal of Social Issues, 29*, 283-300.

Timmins, P., Gallois, C., McCamish, M., & Terry, D. (1998). Sources of information about HIV/AIDS and perceived risk of infection among heterosexual young adults: 1989 and 1994. *Australian Journal of Social Issues, 33*, 179-198.

Timmins, P., Gallois, C., Terry, D., McCamish, M., & Kashima, Y. (1993). Theory of reasoned action and the role of perceived risk in the study of safer sex. In D. Terry, C. Gallois, & M. McCamish (Eds.), *The theory of reasoned action: Its application to AIDS-preventive behaviour* (pp. 153-168). Oxford: Pergamon.

Turner, J. C. (Ed.). (1987). *Rediscovering the social group: A self-categorization theory*. New York: Basil Blackwell.

Weinstein, N. D. (1989). Perceptions of personal susceptibility to harm. In V. M. Mays, G. W. Albee, & S. F. Schneider (Eds.), *Primary prevention of AIDS: Psychological approaches*. Newbury Park, CA: Sage.

Wierzbicka, A. (1995). Everyday conceptions of emotion: A semantic perspective. In J. A. Russell, J.-M. Fernández-Dols, A. S. R. Manstead, & J. C. Wellenkamp, (Eds.), *Everyday conceptions of emotion: An introduction to the psychology, anthropology and linguistics of emotion* (pp. 17-48). London: Kluwer.

– 11 –

Rewards and Risks of Exploring Negative Emotion: An Assimilation Model Account

Lara Honos-Webb[a], Linda M. Endres[b],
Ayesha Shaikh[c], Elizabeth A. Harrick[d],
James A. Lani[b], Lynne M. Knobloch-Fedders[c],
Michael Surko[e], William B. Stiles[b]

[a]Santa Clara University, [b]Miami University
[c]Northwestern University, [d]Earlham College,
[e]Mount Sinai Adolescent Health Center

"I am angry at my father. It took me five years of therapy to say that."

—Richard Gere's character in the movie Pretty Woman
(Milchan & Marshall, 1990)

Psychotherapy clients often believe that the verbal expression of negative emotion will disrupt their daily lives. They are afraid that if they begin to express painful feelings, the pain will never stop. Many also express the concern that talking about traumatic experiences would make the events and the attendant feelings "real." They insist that they have sought therapy in order to feel better, not worse. However, despite this reluctance, therapists frequently encourage clients to verbally express their most painful emotions. Such expression is central to many theories of psychotherapy including psychodynamic (Strupp & Binder, 1984), Gestalt (Perls, 1969), and more recent experiential psychotherapies (e.g., Greenberg, Rice, & Elliott, 1993). In this chapter we present three studies that used the Assimilation Model (Stiles, Elliott, Llewelyn, Firth-Cozens, Margison, Shapiro, & Hardy, 1990) to explore the role of emotional expression in the therapeutic process. We found that under certain circumstances verbal expression of negative emotion seemed to facilitate healing, as therapists insist, but, in other situations, clients may be justified in their fear of putting their painful experiences into words.

The first two studies we review in this chapter tracked the assimilation process in psychotherapy cases. They illustrate how the intensification of negative emotion in the context of supportive therapy can lead to long-term psychological symptom reduction. The third study we review is a laboratory study that explored the relationship between assimilation and physical health. It highlights the potential for harm in giving voice to problematic experiences without some supportive context (such as therapy) to help contain and fully integrate them.

THE ASSIMILATION MODEL

The assimilation model is a metatheoretical model of the cognitive and affective bases of the therapeutic process (Honos-Webb & Stiles, 1998; Stiles et al., 1990). It outlines a systematic series of changes in a client's representation of problematic experiences (such as threatening or painful thoughts, emotions, or memories) from unacknowledged and disowned, to accepted, integrated, and finally mastered. The assimilation approach has been applied to psychodynamic (Stiles, Barkham, Shapiro, & Firth-Cozens, 1992), cognitive-behavioral (Stiles et al., 1990; Stiles, Morrison, Haw, Harper, Shapiro, & Firth-Cozens, 1991), and experiential psychotherapies (Honos-Webb, Stiles, Greenberg, & Goldman, 1998).

There are alternative, complementary formulations of the assimilation model. In the schema formulation (Stiles et al., 1990), therapeutic progress is defined as the process of a client's assimilation of problematic experiences into *schemas* developed in the therapeutic interaction. Schemas in this model are defined as ways of organizing thought, feeling, and action. Experiences are problematic when they conflict with existing core schemas and cannot be meaningfully integrated into the individual's way of cognitively and affectively organizing their world. Problematic experiences challenge assumptions on which these dominant schemas rest. Assimilation is the process by which the core schema slowly changes to accommodate the problematic experience. Simultaneously, the individual's way of thinking about the problematic experience changes to facilitate assimilation into a schema. The schema will now be able to explain, embrace, or include the previously problematic experience.

In the voices formulation of the assimilation model (Honos-Webb & Stiles, 1998), all experiences, including problematic experiences, are conceptualized as separate, active *voices* within the individual rather than as abstract memories or passive stores of information. Each voice is composed of traces of similar or compatible experiences and has its own feelings, motives, and informational content. Individual voices are linked by *meaning bridges* to form the self or personality. A meaning bridge is an understanding and open communication between two or more voices, often resulting in a close bond in which individual voices that were once at odds with one another have softened in order to peacefully coexist. The voices of problematic experiences, however, are often ignored and have few, if any, meaning bridges within the dominant community of voices in the personality. Assimilation in this context is the process of nurturing communication between the dominant voices and the voice of the problematic experience in the service of creating a meaning bridge. This formulation views each individual as a community of voices or a "community of selves" (Mair, 1977), rather than a monolithic Self. A person's experience of self at any moment is the result of the dialogue among the salient voices of experience. This construction of the self is compatible, though not identical, with post-modern theories that reject a stable, autonomous, unitary self in favor of context-dependent, shifting, multiple selves (e.g., Bogart, 1994; Hermans, Kempen, & van Loon, 1992; Hillman, 1975; Mair, 1977; Morawski, 1994; Riger, 1992; Watkins, 1986).

The sequence of assimilation stages or levels, incorporated into the Assimilation of Problematic Experiences Scale (APES; see Table 11.1), represents a changing relationship between two voices—a dominant voice representing the community and the unwanted voice, or problematic experience—or, in the schema formulation, the individual's changing experience of the problem. At stage 0, warded off, the problematic experience is outside the client's awareness. The client either does not remember the experience or does not construe it as problematic. The feelings associated with this stage are neutral. At stage 1, unwanted thoughts, the therapist may point out the problematic experience or external circumstances may make it salient, but the client tries to avoid the issue. Affect around the problematic experience is negative (perhaps even irritable and uncomfortable), but the client still does not acknowledge the problem.

By stage 2, vague awareness/emergence, the client declares that a specific area or theme is problematic, but he or she is unable to pinpoint the problem. In other words, the problematic voice enters the client's awareness, and this new awareness is often experienced as painful, disorienting, and anxiety provoking. It is at this stage that the client experiences the most painful affect. As the new voice becomes stronger, a specific voice that was previously just a member of the dominant community becomes identifiable and separable as it emerges to oppose the problematic voice and maintain the status quo in the community.

The problematic voice positions itself against this dominant voice resulting in problem statement or clarification, stage 3. The client is now able to clearly and specifically state a problem on which to work. Although affect is still negative at stage 3, it is no longer overwhelming. At stage 4, insight, the client is able to generate some theory of the problem's cause or otherwise connect it to

TABLE 11.1

Assimilation of Problematic Experiences Scale (APES)

0. Warded off.	Content is unformed; client is unaware of the problem. An experience is considered warded off if there is evidence of actively avoiding emotionally disturbing topics (e.g., immediately changing subject raised by the therapist). Affect may be minimal at stage 0, reflecting successful avoidance; vague negative affect (especially anxiety) is associated with stages 0.1 to 0.9.
1. Unwanted thoughts.	Content reflects emergence of thoughts associated with discomfort. Client prefers not to think about it; topics are raised by therapist or external circumstances. Affect is often more salient than the content and involves strong negative feelings - anxiety, fear, anger, sadness. Despite the feelings' intensity, they may be unfocused and their connection with the content may be unclear. Stages 1.1 to 1.9 reflect increasingly stronger affect and less successful avoidance.
2. Vague awareness/ emergence.	Client acknowledges the existence of a problematic experience, and describes uncomfortable associated thoughts, but cannot formulate the problem clearly. Affect includes acute psychological pain or panic associated with the problematic thoughts and experiences. Stages 2.1 to 2.9 reflect increasing clarity of the experience's content and decreasing intensity and diffusion of affect.
3. Problem statement/ clarification.	Content includes a clear statement of a problem—something that could be worked on. Affect is negative but manageable, not panicky. Stages 3.1 to 3.9 reflect active, focused work toward understanding the problematic experience.
4. Understanding/ insight.	The problematic experience is placed into a schema, formulated, understood, with clear connective links. Affect may be mixed, with some unpleasant recognitions, but with curiosity or even pleasant surprise of the "aha" sort. Stages 4.1 to 4.9 reflect progressively greater clarity or generality of the understanding, usually associated with increasingly positive (or decreasingly negative) affect.
5. Application/ working-through.	The understanding is used to work on a problem; there is reference to specific problem-solving efforts, though without complete success. Client may describe considering alternatives or systematically selecting courses of action. Affective tone is positive, businesslike, optimistic. Stages 5.1 to 5.9 reflect tangible progress toward solutions of problems in daily living.
6. Problem solution.	Client achieves a successful solution for a specific problem. Affect is positive, satisfied, proud of accomplishment. Stages 6.1 to 6.9 reflect generalizing the solution to other problems and building the solutions into usual or habitual patterns of behavior. As the problem recedes, affect becomes more neutral.
7. Mastery.	Client successfully uses solutions in new situations; this generalizing is largely automatic, not salient. Affect is positive when the topic is raised, but otherwise neutral (i.e., this is no longer something to get excited about).

other areas in his or her life. In terms of the voices metaphor, the previously warded-off or unwanted voice now forms a meaning bridge with the dominant voice, resulting in mutual understanding or accommodation (*softening*), such that the new voice joins the established community as an accepted part of the self. Both the dominant and problematic voices begin to express new sentiments (such as interest and wonder) in the process of constructing this meaning bridge.

In the later stages of the assimilation model, the insights gained through assimilation of the problematic experience are applied to behavior outside of therapy. At stage 5, application/working through, the client tries different behaviors without specific success. The client achieves a solution to a specific problem at stage 6, problem solution, and is often proud of this accomplishment. At stage 7, mastery, the new interpersonal and intrapersonal behaviors are applied to various new situations, and as such they become an accepted and unremarkable part of the client's repertoire. The 0 to 7 APES is conceived of as a continuum, not as discrete stages, so intermediate stages are possible (e.g., 2.5 represents a stage midway between vague awareness/emergence and problem statement/clarification). Figure 11.1 is a visual representation of the degree of positive and negative affect theoretically experienced at each APES stage along with the salience, or amount of attention, the client directs to the problematic experience.

To summarize, in the schema formulation of the assimilation model, problematic experiences slowly and painfully emerge into the individual's awareness. Through repeated attempts at verbal expression, the problem eventually becomes clearly formulated before being resolved, and ultimately contributes to the individual's growth. The four central concepts in the voices formulation of the assimilation process are: (a) a dominant community of voices representing accepted experiences, (b) the voice of the problematic experience (the underdog voice), (c) a voice that emerges from the community to challenge the unwanted experience (the dominant, or top dog voice), and (d) a meaning bridge that links the previously unwanted voice to the dominant community of voices.

The two formulations of the model employ slightly different definitions of the term schema. In the schema formulation, the term does not distinguish between what might be called the established community of voices (in other models, the self) and the cognitive structures (attitudes, beliefs, concepts) used to bind the community together and form links with the problematic experience. In the voices formulation, the term "schema" is used only for the latter—the meaning bridge that is constructed between the dominant and problematic voices, which allows them to understand each other, and the links that bind the community of voices together. Despite these different conceptualizations or

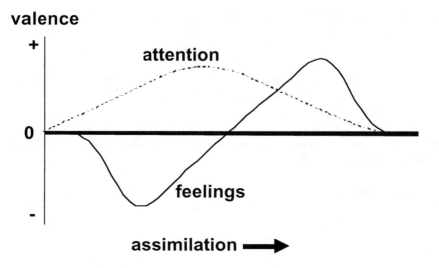

FIG. 11.1. Theoretical relations among assimilation, salience, and affect regarding a problematic experience.

metaphors, however, the two formulations of the assimilation model ultimately describe the same process—the integration of a problematic experience into the individual's personality (Stiles, Honos-Webb, & Lani, 1999).

Next, we review two assimilation analyses of psychotherapy cases, which illustrate applications from both formulations: the first case example (Vicky; Knobloch, Endres, Stiles, & Silberschatz, 2001) employed the schema formulation, and the second case example (Millie; Lani, Shaikh, Stiles, & Silberschatz, 1998) used the voices formulation. These two examples from psychotherapy are followed by a review of an application of the assimilation model to a laboratory writing study (Honos-Webb, Harrick, Stiles, & Park, 2000), based on a paradigm developed by Pennebaker (1993).

THE CASE OF VICKY

As described earlier, the assimilation model hypothesizes that the verbal expression of emotion facilitates the cognitive and affective assimilation of problematic experiences into existing schemas or into schemas developed in the therapeutic interaction. For the purposes of demonstration, we present some key passages taken from the transcripts of one woman's therapy, Vicky (a pseudonym). In doing so, we focus on Vicky's struggle to overcome her sexual anxieties. Although this is only one of the major issues Vicky addressed during the course of therapy, it is representative of successful assimilation and fits within the scope of this chapter.

Vicky was a 20-year-old university student who presented in therapy with the goals of becoming more relaxed with her sexuality, improving her relationship with her parents, and finding some career direction. More specifically, Vicky complained that she was unable to respond sexually in her encounters with men and was afraid she might be "frigid." Vicky also reported difficulty studying and was unsure about her future career plans. Finally, Vicky alluded to stormy, and sometimes violent, relationships with her parents. English was Vicky's second language (Spanish was her first). She had grown up primarily in Central and South America and had studied in Europe before moving to the United States.

Vicky was a client of the Mt. Zion Brief Therapy Research Project and her therapy has been studied previously (Knobloch, Endres, Stiles, & Silberschatz, 2001; Norville, Sampson, & Weiss, 1996). She completed 18 therapy sessions as well as an intake assessment interview, post-therapy evaluation, and an 18-month follow-up interview.

Before therapy began, Vicky completed the SCL-90 (Derogatis, Lipman, & Covi, 1973) and wrote a detailed list of her complaints and her goals for therapy. The SCL-90 is a self-report symptom checklist that assesses psychological distress and psychopathology. The scale includes nine primary symptom dimensions and three global indices of distress. Each of the 90 items is rated on a 5-point scale of distress (0-4) from "not at all" to "extremely." By the post-therapy evaluation, Vicky's overall SCL-90 score had dropped from 1.17 to .72. Additionally, according to the client's, therapist's, and outside evaluator's ratings, Vicky resolved a significant number of her complaints and accomplished a significant number of her goals.

Vicky began her intake session by telling the therapist about her sexual anxieties, which she believed prevented her from responding to men, both emotionally and physically. The following excerpt is how Vicky introduced her concern in therapy:

> V: ...I think the main things that I would like to sort of get over are insecurity and that insecurity is bothering me especially in the sphere of my sexuality because I find that every time I am going to uh, uh—I'm starting a relationship with somebody or somebody asks me out or something, that—I am immediately apprehensive about the sexual part and I'm so scared that I feel that interferes a lot with my emotional—with my emotional or sexual response and that is the main thing that is causing me trouble right now that uh, it's making me feel quite anxious—that I—I am scared of actually having sexual intercourse with anybody because I'm also scared of being seen as somebody who is frigid... [*Intake session; stage 3*]

Vicky's affect in telling the therapist about this problem was not overwhelming, but was profoundly negative as illustrated by her use of such terms as "apprehensive," "insecure," "bothering," "scared," and "anxious." Based on the APES scale, we rated this passage as stage 3, problem statement, because Vicky was plainly stating the nature of her problem. She knew what the problem was—that the mere idea of sex distressed her—but she didn't provide any context for the problem, such as what might have caused it. In other words, her distress around sexuality was unconnected to any schemas.

It is important to note that not all stages hypothesized by the APES scale must be represented in therapy. Assimilation is not limited to therapy and occurs naturally before therapy begins, between therapy sessions, and even after therapy is terminated (as we illustrate later). In Vicky's case, the theme of her sexual distress began in therapy with this excerpt at stage 3, problem statement.

In therapy Vicky also often spoke of her troubled relationship with her mother. She told stories of times, beginning when she was 12 or 13 years old, when her mother called her a whore (in Spanish, a *calentona*) just for looking at a boy. The therapist saw a connection between these epithets and Vicky's current anxieties around sexuality, and even offered her the insights, but Vicky generally ignored these comments or denied the connection. It wasn't until session 12 that Vicky made the connection herself.

T: And of course you have the belief that your mother slept with your father and never enjoyed it the way a whore would do.
V: That's interesting, because then in order to avoid being a whore then I should prevent myself from having sex too.
T: Yeah, that seems it might fit together. [*Session 12; stage 4*]

Here, the emotion changed. Vicky no longer spoke of feeling anxious, uncomfortable, or embarrassed; instead, the distinguishing characteristic of this passage was the transition to a more neutral emotion—interest. The other feature that marked this excerpt as a moment of insight, stage 4, was that Vicky herself connected her desire to not be seen as a whore with her ambivalence about sex. Although the therapist had previously mentioned this connection to Vicky a few times, we did not consider it an insight event until the client herself connected the two ideas.

In later sessions, Vicky still struggled with her fear of being perceived as a whore, but the focus changed from "Why do I have this problem? What is it all about?" to "How can I overcome it?" Vicky's affect in these later stages was no longer as negative as in earlier sessions, but was more matter-of-fact and neutral.

T: Well the problem for you isn't it is can you possibly let go and become really excited—

V: And not be a whore.

T: —and deeply and maintain your own sense of dignity and worthwhileness.

V: Yeah. [*Session 19; stage 4*]

We rated this passage again as stage 4, but note how Vicky can say what was for her a very loaded word, "whore," without the familiar accompanying discomfort.

Vicky terminated therapy soon after this exchange with her therapist. Her work in therapy ended with symptomatic improvement, as indicated by the drop in her SCL-90 score from 1.17 at intake to .72 at termination. Although the SCL-90 change revealed a decrease in Vicky's symptomatic complaints, her score at termination was still rather high. We believe that Vicky's incomplete symptomatic improvement as measured by the SCL-90 reflected her minimal progress on other themes not discussed in this chapter (see Knobloch et al., 2001, for a more extensive discussion).

Although we do not have follow-up questionnaire data, there was some qualitative evidence that assimilation continued. At the follow-up session, which occurred 18 months after her final therapy session, Vicky recounted a successful solution, stage 6, to her sexual anxiety problem. Her mood in talking about her sexuality was now marked by exuberance and mastery:

V: When I was, I was in New York last summer and I was with my parents and uhm, my father had a friend there and I started going out with him and—it was basically my idea to uh, you know, go to his apartment and and have sex, and it wasn't—it went very fluidly, it was—I expected that and he expected that, and I was actually feeling very exuberant af—about it afterwards, because I felt like I was totally in control of what was going on—

T: Mm-hmm.

V: And I was not waiting to see what happened, I was actually directing the whole thing. Uhm, so it—in that sense it's changed and I'm, I'm very glad about that. [*Follow-up session; stage 6*]

Thus, Vicky conveyed a clear sense of having solved this problem, though her having sex with a friend of her father's might be considered noteworthy or potentially problematic.

Vicky's therapy illustrates the idea that emotional expression in psychotherapy needs to be connected to the themes of problematic experiences, not just expressed in an unconnected fashion. Vicky spent the first 12 sessions in therapy primarily expressing her negative emotions, but these emotions remained fundamentally disconnected to any insights about her difficulties with sexuality. It may be that expressing negative emotion in psychotherapy leads to positive outcomes if it is analyzed, integrated, and understood, but poorer outcomes if it remains unconnected to schematic assimilation of problematic experiences.

THE CASE OF MILLIE

In the case of Millie (a pseudonym), we used the voices formulation of the assimilation model to track her progress through therapy. According to our analysis, her emerging "Independence and Autonomy" voice was challenged, and initially suppressed, by her "Please Everyone, Don't-Rock-the-Boat" voice. Within a matter of weeks, however, the Independence voice fought successfully to be heard before finally assimilating into Millie's community of voices.

Millie was a 54-year-old woman also seen in the Mt. Zion Brief Psychotherapy Research Project (Norville et al., 1996) and her therapy had been studied earlier (Curtis & Silberschatz, 1997; Lani, Shaikh, Stiles, & Silberschatz, 1998, Lani, Stiles, Shaikh, & Silberschatz, 1998). She was married and had two adult children. At the time she entered therapy, her husband was suffering from a serious degenerative illness that had left him deaf, mute, and permanently hospitalized. As a result, Millie became the household wage earner and was thrust into a job market for which she felt ill-prepared. An employment counselor referred Millie to therapy because she appeared to have unresolved therapeutic issues that interfered with her ability to hold a job. At intake, Millie described feeling "terribly out of it" and lonely. She was afraid of what her future would hold and felt she lacked direction and purpose in life. She also described feeling "uncomfortable in [her] own skin." During the course of therapy, Millie related a history of always trying to please the significant people in her life so that they would love her. She felt controlled by them, and had never allowed herself to act on her own desires.

In 16 psychodynamic psychotherapy sessions, Millie explored her experiences as a sister, wife, and daughter, and examined feelings of anger and sadness that stemmed from childhood experiences. Given the change in her SCL-90 scores (pre-therapy = .67 to post-therapy = .18), we considered this a successful case.

We labeled the major theme in Millie's therapy "control," which we conceptualized as a dialogue between an *underdog* voice that rallied for Millie's autonomy and a *top dog* voice that suppressed her independence. With the help of therapy, Millie was able to express the underdog voice and understand how the top dog voice functioned to control her. In the following excerpts, we show how Millie progressed from attempting to avoid her underdog voice and associated painful emotions (stages 0 and 1), through the emergence of an angry underdog voice (stages 2 and 3), and finally to the joining of these two positions into a community of voices that had greater understanding of both the top dog and underdog positions (stage 4).

In the first session, Millie warded off her feelings of anger at being controlled. In the following passage, which we rated as a late stage 0, she described these feelings with considerable distance:

M: My life has to go someplace, and it's just not—I'm in a holding pattern and I'm unwilling to break out of it. And there is I'm sure some anger—usually when people are in this kind of blah state, which really borders on depression, they're angry. I mean, I know enough about the theory of psychology to know that. [*Session 1; Stage 0.8*]

Although Millie described feeling like she was in a "holding pattern," she did not yet articulate a clear problem on which to work. She expressed distress (as evidenced by her reference to feeling "blah" and "angry") but held those feelings at bay by talking about them in the third person and thereby depersonalizing them.

Nevertheless, by continuing to talk about her feelings, Millie's problems became more focused. Although she started with vague allusions to problematic experiences, she quickly progressed to owning her anger and describing the specific situation that instigated it—being pushed around.

M: I think my anger, to some extent, is legitimate—because I'm in an uncomfortable position, and people tend to get angry when they're pushed in a way that they don't wanna be pushed. On the other hand it's self-defeating, because I am in the circumstance in which I am. Uh—there isn't gonna be a knight on a white steed coming and picking me up and saying, "All right, sweet Millie, you don't have to do anything; I'll do it all for you." Those are the realities, and uhm, if I'm not dealing with them, I certainly ought to be. [*Session 1; Stage 1.8*]

As an example of late stage 1, unwanted thoughts, this passage shows Millie's underdog voice avoiding her feelings of anger at being pushed around ("people tend to get angry...") whereas her top dog voice was distinctly sarcastic ("All right, sweet Millie..."). The underdog voice again used the third person to distance itself from the painful emotion, and the top dog voice reacted with quick sarcasm to discredit and suppress the underdog's observations. In this passage, the top dog voice was still in control, but was threatened by the underdog.

In the second session, Millie's underdog voice remained tentative. As she talked about a lifetime of dependency on others, she felt lonely and isolated. The underdog voice emerged slowly, and ultimately challenged the top dog voice.

M: I've worked a lifetime at eliminating my own desires [*crying*] in favor of somebody else's and now I'm told, "What are you moping about?" You know—and I wanna say, "Hey, wait a minute." This is what I was taught to be, and now it's no good anymore. It takes so much time to even know what you WANT when you have really never had the opportunity to express it. When you took your cue from what this one wanted, and what that one wanted, and the third one needed. And you taught yourself not to be disappointed year after year because you couldn't really do what you wanted to do. I've lost the capacity now, to really know what I want. I don't really know what I want. [*Session 2; Stage 2.5*]

APES stage 2 is characterized by an emerging underdog voice accompanied by intense negative affect. Millie experienced this as grief at having her own desires suppressed by others for so long that she now didn't know what she wanted. While the underdog voice cried, "I've worked a lifetime at eliminating my own desires," the top dog voice seemed to belittle it by saying, "What are you moping about?" The underdog voice defensively responded with, "Hey wait a minute... This is what I've been taught to be," but ultimately lost its resolve in the face of the disapproving top dog voice and ended with, "I've lost the capacity...I don't really know what I want." This type of sparring relationship between the top dog and underdog voices is a good illustration of APES stage 2.

In the third session, Millie reflected more deeply on her feelings of being controlled. Her affect became less negative as she moved toward formulating a clear problem statement, stage 3.

> T: Is it the same kind of argument that your father, for example, is going to be very hurt if you go out and live your own life, or your husband would be very upset if you had fun?
>
> M: There is definitely is a connection. I FEEL a connection; there is a connection between those things. I definitely feel a connection between those things. What I hear is that I'm not allowed—I'm not allowing myself to be who I am. [Pause] It never was all right to be who I was. But it goes SO FAR back. My God, it goes so far back. It goes back to my earliest memories as a kid, when I was maybe five years old, or even less perhaps, which are my earliest conscious memories. And one thread that runs through my whole life is always being held in check, always being held back—when [pause] what I really would've liked is to have somebody say, "Go out there and do whatever it is and you know we love you enough so you always can come back." And I was given to understand that I wouldn't be loved anymore if I did things that were "wrong" to their way of thinking, and so I adapted, but it's just like when you try to make somebody write with their right hand who is basically left-handed. Something gets thwarted. And I think that's what happened to me—WAY back when. When those qualities that should have been—that I should have been lauded for—were put down as being troublesome and forward, and—I, who am so timid today, uhm, was thought of as being too assertive as a child. [*Session 3; stage 3.5*]

This passage exemplifies APES stage 3, problem statement, because Millie was now able to reflect on both the top dog and underdog perspectives equally. The underdog position was clearly articulated in her descriptions of her childhood. She even told the top dog voice what she would have liked—for someone to say "Go out there and do whatever it is and you know we love you enough so you can always come back." Millie also expressed an understanding of and even compassion for the top dog position in her description of how she adapted for the sake of being loved. This emerging dialogue between the underdog and top dog perspectives shed some light on why Millie had put the interests of others ahead of her own. She *felt* a connection but wasn't able to fully verbalize the connection; therefore we rated the passage as approaching insight, APES stage 3.5.

By the 16th session, Millie talked about allowing herself to become independent enough to determine what she did and did not want, and then to act on it. Although she had effectively pieced together the factors that had contributed to her current problems, she had not yet determined what she would do about them. However, she did recognize that any solution would involve more focused attention on her own desires and needs.

> M: [Sniff] I always have to clear everything with my conscience and a conscience is just an accumulation of things that have been told me over the years. Is it really my conscience, or is it my parents' conscience? Which was good for their generation, but doesn't have to be for me, cannot be for me. They don't live now; if they did they would have to revise their standards, other values. [Pause] And I think that's what I'm still grappling with. All this stuff that's just sitting there, that's telling me how to behave, how to think, TREMENDOUS amount of do's and don'ts.
>
> T: Rather than just letting yourself go ahead and do it?
>
> M: Rather than letting me find out what it is I really am, and wanna do, and then going ahead and doing it. Because I'm [pause] I have seldom been what I really wanted to be. Mostly I've been what I felt people around me wanted me to be. Or thought I was, and then I'd live up to that expectation or try. And when I failed, I put myself down for it, rather than saying, "It's their failure for wanting me to be something that I'm not, and if any failure is there at all, it's because I'm trying to be something that I'm not." But the assertiveness to say, "Hey, that's not my way." This I have never learned. I don't know what that means, how to do that, how to be secure enough to say, "I don't wanna do that."
>
> T: Yet it sounds like you just told me how to do it. [*Session 16; Stage 4.8*]

This passage is an example of a late stage 4, insight. Millie advanced her understanding of the factors that shaped the person she was and, without realizing it, even presented a possible solution to her problem. The top dog and underdog voices alternated in discussing Millie's situation. The top dog talked about "all that stuff that's just sitting there, that's telling me how to behave, how to think, TREMENDOUS amount of do's and don'ts," and the underdog finished the thought with, "... rather than letting me find out what it is I really am, and wanna do and going ahead and doing it." The underdog voice, which had previously verbalized sadness at not knowing what she wanted and having been held back by others, occupied most of the dialogue. Through the development of an understanding between the two voices, Millie was almost prepared to begin work on problem solutions (APES stage 5). Although Millie's affect in this passage was still somewhat negative, we can see how by this last therapy session the transformation of her anger into assertiveness had begun.

Millie's case illustrates how assimilation can be conceptualized as an advancing dialogue between voices. In this view, as with the schema conceptualization, the verbal expression of negative emotion in psychotherapy is key in promoting the early stages of constructive psychological change. As Millie's underdog voice emerged and was acknowledged and understood by her top dog voice, Millie developed insight into her problems. The expression of negative emotion, particularly from the previously marginalized underdog voice, facilitated these insights (including her resolution to become more assertive.)

ASSIMILATION AND HEALTH OUTCOMES

The two psychotherapy case studies we presented speak to the effects of assimilation on psychological health, but do not address its effects on physical health. Recent investigations of assimilation and its effects on physical health have been conducted in the context of what we will call the "Pennebaker paradigm" (Honos-Webb, Harrick, Stiles, & Park, 2000). The Pennebaker paradigm refers to a laboratory procedure in which students are instructed to write for 20 minutes on each of 4 consecutive days about their most traumatic experience. Study participants assigned to this procedure, as compared with those asked to write about innocuous topics, have shown improved immune functioning (Pennebaker, Kiecolt-Glaser, & Glaser, 1988), better self-reported health (Pennebaker & Beall, 1986), and reduced use of health services (Pennebaker & Beall, 1986). Many variations of this procedure have been implemented that replicate the result that it benefits a significant proportion of those who participate (Pennebaker, 1993, 1997).

Although most participants in the treatment (cathartic writing) condition show improvement in health outcomes, past studies have shown that there is substantial variability (Pennebaker, 1993). In an experiment conducted at a Midwestern university in the early spring of 1996, assimilation ratings were given to the four essays written by each of the 43 participants in the cathartic writing condition (one essay a day for 4 days). Both the overall change in assimilation and the highest stage of assimilation reached across the 4 days were correlated with the number of visits students made to the university health center. Visits made the same semester as the experiment represented short-term physical health outcomes, and visits made the following autumn represented long-term physical health outcomes. Assimilation was found to be significantly related to short-term physical health outcomes but not to long-term health outcomes (Honos-Webb et al., 2000).

As shown in Table 11.2, there were significant correlations between assimilation and frequency of health center visits. Both the amount of increase in assimilation and the highest assimilation stage reached were positively correlated with the number of health center visits in the spring of 1996 (short-term health outcome). That is, participants who showed the most progress in assimilation and those who reached the highest assimilation stages had the most health center visits (or negative health impact) in the short term.

TABLE 11.2

Mean Number of Students Visits to the Health Center for Accident and Injury in Each Semester and Correlations With Assimilation Indices

Semester	Students	Visits		Correlation with assimilation	
		Mean	SD	Change	Highest Stage
Autumn 1995	43	.93	1.47	.092	-.026
Spring 1996	43	1.44	2.36	.397**	.304*
Autumn 1996	43	.56	.93	.292	.034

Note. Health center visit frequencies include accident- or injury-related visits only.
*$p < .05$ (2-tailed).
**$p < .01$ (2-tailed).

We suggested two possible explanations for the observed correlation between change in assimilation stage and health center visit frequency (Honos-Webb et al., 2000). One possible explanation for this finding is that there was a restricted range in assimilation ratings for this sample (no participant progressed beyond APES stage 4). Because these early assimilation stages are associated with strong negative emotion, and because negative emotion is associated with increased illness behaviors (Leventhal, Hansell, Diefenbach, Leventhal, & Glass, 1996; VanderVoort, 1995), assimilation at the higher stages (APES stages 5-7), which are associated with neutral and positive affect (see Figure 11.1), may be needed for positive health outcomes. If stages 5 to 7 had been present in the sample, the relationship between assimilation and physical health might have been curvilinear.

Another plausible explanation is that participants who showed the most change in assimilation stage uncovered traumatic experiences that had previously been successfully warded off. As this was just a 4-day writing experiment, they might not have been given enough time or support to resolve and fully integrate the previously unwanted material. The assimilation model suggests that progress across the early stages of assimilation is associated with increasing distress (see Honos-Webb et al., 1998, for a case example). Participants in the Pennebaker paradigm who actively face their traumas but do not resolve them may find this a disruptive and stressful experience, which is subsequently reflected in more health center visits.

To illustrate, we present the final day's writing sample for one participant in this study (Honos-Webb et al., 2000). This writing sample showed a change in assimilation from stage 0 to stage 3, and was therefore among the highest on both assimilation indices (change = 3 and highest = 3). In the semester immediately prior to the experiment, she did not visit the health center. During the short-term follow-up period she made seven visits to the health center, and during the long-term follow-up period she made an additional four visits.

The theme tracked in this sample was "problematic sexual experience." From an assimilation of voices perspective, the underdog's claim of incestuous sexual abuse was minimized by a top dog voice that questioned the incident's importance (and even reality).

For the first 3 days, the participant wrote about a problem she was having with a friend, a topic unrelated to the problematic sexual experience theme. We rated these three essays APES stage 0, warded off. The participant introduced the problematic sexual experience theme on the fourth day.

At the beginning of this fourth essay, she said that her feelings weren't so traumatic as previously and that things with her friend were as they had been before the problem. That is, she began by writing about the less serious problem she had written about on the previous 3 days. As assimilation ratings are made in the context of a specific theme or problem, and this paragraph was not related to the problematic sexual experience, we rated it as stage 0, warded off. By focusing on a minor problem in a current friendship, the participant initially warded off the more painful memories of abuse.

She then described an incident that had occurred 10 years earlier, when she was 9 years old. She began by describing the physical realities of the experience, an incident in which her father invited her into his bedroom because he wanted to talk, as opposed to discussing the emotional reactions and psychological consequences of the event. In this part of the essay, the problematic experience appeared to be at stage 1, unwanted thoughts. She then described how her father had asked her to massage him sexually, which she had done. She described how she felt upset and uncomfortable about doing this. It appeared that at this point she fully accepted the validity of the traumatic experience, indicated by her wondering if her father remembered committing the abuse (APES stage 2). The abuse had happened only once, though on one subsequent occasion, her father had asked for another massage as a condition for taking her to a sports event and she had opted instead not to go. Near the end of her writing, she suggested that, as a consequence of this episode, she had not been able to have a sexual relationship with anyone, and that she sometimes felt ill when she looked at her father. Yet in the same passage, a seemingly opposing voice minimized the incident's importance and even questioned whether it really happened.

The participant's admission that she had not been able to have a sexual relationship represented a traditional problem statement, APES stage 3, insofar as she relayed a specific problem that could be worked on. Both voices were present throughout this passage with approximately equal weighting. The participant acknowledged a current problem, sexual dysfunction (underdog), but simultaneously minimized its importance (top dog). At the end of the essay, both voices remained active as the underdog voice expressed disgust at her father, whereas the top dog expressed doubt as to the validity of the story.

This example illustrates the possibility that emergence of problematic or traumatic material in the absence of other supports may be distressing and may in turn lead to negative health outcomes. Though not explicitly stated, it appeared to us that this participant had not previously described this incident to anyone. The trauma-writing task seemed to have reminded her of this story, and after 3 days of warding it off, she gave in and gave vent to it. We speculated that the retraumatization of explicitly reviewing this incident may have raised her level of stress. The well-documented relationship between stress and immunosuppression and/or other negative health impacts (Maier, Watkins, & Fleshner, 1994) may explain the relationship between progress in assimilation and number of health center visits.

CONCLUSIONS: IMPLICATIONS FOR EMOTIONAL EXPRESSION

The collection of studies presented offers different conclusions about the effects of the verbal expression of emotion. The first two qualitative assimilation analyses suggest that in the context of supportive psychotherapy, the expression of negative emotion can lead to symptom reduction. More specifically, Vicky's case suggests that the verbal expression of negative emotion is particularly helpful when the client makes meaning out of the negativity in the context of an insight event. Millie's case analysis converged on this finding by suggesting that it was the transformation of her newly assimilated anger into assertiveness that promoted symptom reduction.

Theoretically, the expression of negative affect predominates at stage 2, vague awareness/emergence. These analyses suggest that simply expressing negative emotion as catharsis, outside of a meaning-making context, is not by itself sufficient to promote therapeutic change. Rather, the expression of negative affect was most helpful to these clients in moving them through the very early stages of assimilation, before meaning-making could occur (in other words, before the experience was incorporated into a schema or before the underdog voice was made part of the community of voices).

Curiously, our studies also seem to contradict each other. In Millie's case, assimilation reached a high of nearly stage 5 and was associated with near complete remission of psychological symptoms for her. In Vicky's case, assimilation at termination reached a high of stage 4 and was associated with moderate symptom reduction as measured by the SCL-90 (the passage rated stage 6 was from the follow-up session 18 months later). In the laboratory study, however, assimilation up to stage 4 was associated with poorer health outcomes. These latter findings are consistent with the Honos-Webb et al. (1998) study, which found no improvement in measures of psychological symptomatology for assimilation below stage 4 in a two-case comparison. It may be that the transition between insight, stage 4, and application, stage 5, is crucial for symptom reduction. On the other hand, it may be that the relationship between highest stage of assimilation and negative health outcome in the laboratory study was an artifact of the relationship between change in assimilation and negative health outcome—in other words, those who achieved the highest stages of assimilation were also those who evidenced the greatest change in a very short period of time (4 days). Alternatively, the support and validation psychotherapy clients receive from their therapists may buffer them from the isolating, and potentially overwhelming, effects of delving into traumatic experiences. Future studies need to address this inconsistency and further test the effects of assimilation on health.

This application of the assimilation model to the Pennebaker paradigm also suggests caution in making blanket statements about the benefits of expressing painful emotions related to traumatic events. Our analyses revealed that there were casualties among the participants in this experiment. Although the overall effect of this procedure has generally been positive, the method of aggregating all participants may have hidden a potentially identifiable group of participants who were harmed by the procedure. Clients who showed greater change in assimilation had more immediate negative health impacts.

The studies reviewed in this chapter suggest that expression of negative emotion can lead to both beneficial and disruptive outcomes. The assimilation model provides a theoretical basis for distinguishing between the two. In assimilation terms, the expression of painful emotions without the opportunity to fully integrate them into schemas or "the community of voices" can be disruptive and damaging. However, when a previously unwanted voice or emotional experience is expressed and integrated, the client will likely experience health benefits, alleviation of intrusive symptoms, and a sense of greater control over his or her life. These findings are consistent with findings regarding those who benefit from participation in the Pennebaker paradigm. Content analyses of the writings of participants found that an increase in causal or insight words over the course of writing predicted improved health outcomes (Pennebaker, Mayne, & Francis, 1997). This parallels recent theory that suggests that the expression of traumatic experiences needs to be accompanied by meaning-making coping processes in order to facilitate resolution (Park & Folkman, 1997).

Although we have suggested tentative interpretations for our data, more research needs to be conducted in order to explicate the role of assimilation in psychological and physical health. Specifically, our conclusion, summarized in Table 3, that assimilation up to stage 4 (insight) maintains or promotes symptomatic distress, whereas assimilation that reaches stage 5 (working though) or higher results in improved mental or physical health, needs to be researched.

TABLE 11.3

Summary of Associations Between Assimilation and Health Outcomes

Health Outcome Domain	Change occurred within assimilation stages	
	0-3	4-7
Psychological	No improvement	Improvement
Physical	Negative health outcomes	Unknown

One implication of these results is that unassimilated problems take their toll in the physical health domain as well as the psychological realm. The assimilation model concurs with the suggestion that problematic experiences and voices (stages 0-3) that are not expressed verbally will find their expression in symptoms and other negative health outcomes.

We conclude with the words of another client (not analyzed in this chapter) who dared to explore and express her painful emotions in spite of her initial fears. The following are the final words of her treatment:

C: Well, it felt like you were really, yeah you were really good and very helpful to me. And yeah, I don't know, it just, and a lot of times you said things, and it would be like, "Why did she say this now, she read my mind, she knows." [Laugh] It was almost like, "Oh, I hate her for this one almost, these—

T: These mixed feelings almost . . .

C: Yeah, these mixed feelings. But then like at the same time, it kinda, well it did, put me in touch with these childhood feelings and I think I said this last time, it's kinda really difficult. But first of all, having the willingness and getting back to that and like experiencing this, ummm [sigh] . . .

T: Painful feeling.

C: Yeah, really painful feelings. But, it did help [laugh].

REFERENCES

Bogart, V. (1994). Transcending the dichotomy of either "subpersonalities" or "an integrated unitary self." *Journal of Humanistic Psychology, 34*, 82-89.

Curtis, J. T., & Silberschatz, G. (1997). The plan formulation method. In T. Eells (Ed.), *Handbook of psychotherapy case formulation* (pp.116-136). New York: Guilford Press.

Derogatis, L. R., Lipman, R. S., & Covi, L. (1973). SCL-90: An outpatient psychiatric rating scale—Preliminary report. *Psychopharmacology, 9*, 13-27.

Greenberg, L. S., Rice, L. N., & Elliott, R. (1993). *Facilitating emotional change: The moment-by-moment process.* New York: Guilford Press.

Hermans, H. J. M., Kempen, H. J. G., & van Loon, J. P. (1992). The dialogical self: Beyond individualism and rationalism. *American Psychologist, 47*, 23-33.

Hillman, J. (1975). *Revisioning psychology.* New York: Harper & Row.

Honos-Webb, L., Harrick, E. A., Stiles, W. B., & Park, C. L. (2000). Assimilation of traumatic experiences and physical health outcomes: Cautions for the Pennebaker paradigm. *Psychotherapy, 37*, 307-314.

Honos-Webb, L., & Stiles, W. B. (1998). Reformulation of assimilation analysis in terms of voices. *Psychotherapy, 35*, 23-33.

Honos-Webb, L., Stiles, W. B., Greenberg, L. S., & Goldman, R. (1998). Assimilation analysis of process-experiential psychotherapy: A comparison of two cases. *Psychotherapy Research, 8*, 264-286.

Knobloch, L. M. Endres, L. M., Stiles, W. B., & Silberschatz, G. (2001). Convergence and divergence of themes in successful psychotherapy: An assimilation analysis. *Psychotherapy, 38*, 31-39.

Lani, J. A., Shaikh, A., Stiles, W. B., & Silberschatz, G. (1998). *Millie: An assimilation of voices case study.* Unpublished manuscript, Department of Psychology, Miami University, Oxford, OH.

Lani, J. A., Stiles, W. B., Shaikh, A., & Silberschatz, G. (1998, June). Hearing change in clients' narratives: An assimilation of voices perspective. In G. Silberschatz (Moderator), *Assimilation analyses of short-term psychodynamic therapies.* Panel presented at the Society for Psychotherapy Research meeting, Snowbird, UT.

Leventhal, E. A., Hansell, S., Diefenbach, M., Leventhal, H., & Glass, D. C. (1996). Negative affect and self-report of physical symptoms: Two longitudinal studies of older adults. *Health Psychology, 15*, 193-199.

Maier, S. F., Watkins, L. R., & Fleshner, M. (1994). Psychoneuroimmunology: The interface between behavior, brain, and immunity. *American Psychologist, 49*, 1004-1017.

Mair, M. (1977). The community of self. In D. Bannister (Ed.), *New perspectives in personal construct theory* (pp. 125-149). London: Academic Press.

Morawski, J. G. (1994). *Practicing feminisms, reconstructing psychology: Notes on a liminal science.* Ann Arbor, MI: The University of Michigan Press.

Norville, R., Sampson, H., & Weiss, J. (1996). Accurate interpretations and brief psychotherapy outcome. *Psychotherapy Research, 6*, 16-29.

Milchan, A. (Producer), & Marshall, G. (Director). (1990). *Pretty Woman* [Video]. (Available from Buena Vista Home Video, Dept. D. S., Burbank, CA 91521)

Park, C. L. & Folkman, S. (1997). Meaning in the context of stress and coping. *General Review of Psychology, 1*, 115-144.

Pennebaker, J. W. (1989). Confession, inhibition, and disease. *Advances in Experimental Social Psychology, 22*, 211-244.

Pennebaker, J. W. (1993). Putting stress into words: Health, linguistic and therapeutic implications. *Behavior Research and Therapy, 31*, 539-548.

Pennebaker, J. W. (1997). Writing about emotional experiences as a therapeutic process. *Psychological Science, 8*, 162-166.

Pennebaker, J. W., & Beall, S. K. (1986). Confronting a traumatic event: Toward an understanding of inhibition and disease. *Journal of Abnormal Psychology, 95*, 274-281.

Pennebaker, J. W., Kiecolt-Glaser, J. K., & Glaser, R. (1988). Disclosure of traumas and immune function: Health implications for psychotherapy. *Journal of Consulting and Clinical Psychology, 56*, 239-245.

Pennebaker, J. W., Mayne, T. J., & Francis, M. E. (1997). Linguistic predictors of adaptive bereavement. *Journal of Personality and Social Psychology, 7*, 863-871

Perls, F. S. (1969). *Gestalt therapy verbatim*. Lafayette, CA: Real People Press.

Riger, S. (1992). Epistemological debates, feminist voices: Science, social values, and the study of women. *American Psychologist, 47*, 730-740.

Stiles, W. B., Barkham, M., Shapiro, D. A., & Firth-Cozens, J. (1992). Treatment order and thematic continuity between contrasting psychotherapies: Exploring an implication of the assimilation model. *Psychotherapy Research, 2*, 112-124.

Stiles, W. B., Elliott, R., Llewelyn, S. P., Firth-Cozens, J. A., Margison, F. R., Shapiro, D. A., & Hardy, G. (1990). Assimilation of problematic experiences by clients in psychotherapy. *Psychotherapy, 27*, 411-420.

Stiles, W. B., Honos-Webb, L., & Lani, J. A. (1999). Some functions of narrative in the assimilation of problematic experiences. *Journal of Clinical Psychology, 55*, 1213-1226.

Stiles, W. B., Morrison, L. A., Haw, S. K., Harper, H., Shapiro, D. A., & Firth-Cozens, J. (1991). Longitudinal study of assimilation in exploratory psychotherapy. *Psychotherapy, 28*, 195-206.

Strupp, H. H. & Binder, J. L. (1984). *Psychotherapy in a new key: A guide to time-limited dynamic psychotherapy*. New York: Basic Books.

VanderVoort, D. J. (1995). Depression, anxiety, hostility, and physical health. *Current Psychology: Developmental, Learning, Personality, Social, 14*, 69-82.

Watkins, M. (1986). *Invisible guests: The development of imaginal dialogues*. Hillsdale, NJ: Lawrence Erlbaum Associates.

– 12 –

Blocking Emotions:
The Face of Resistance

Kathleen W. Ferrara
Texas A & M University

INTRODUCTION

This chapter examines the linguistic manifestations of the interactional phenomenon of resistance. Although the task of every client in counseling or psychotherapy is to use everyday language to discuss his or her experiences in as free and open a manner as possible, clients in psychotherapy frequently block emotions and withhold personally revealing material. The motivation for this opposition, according to Freud (1901/1960), Greenson (1967), and Mahalik (1994), is always the avoidance of some painful feeling. Some of the negative emotions include sadness, grief, sense of loss, remorse, shame, anger, hostility, resentment, and guilt. Although one of the greatest problems facing healthcare providers such as doctors or therapists is non-compliance with therapeutic suggestions, the oppositional behavior called resistance is little understood. Not all of it is intrapsychic; some is interactional.

Advances in discourse analysis and recent work on therapeutic discourse (Ferrara, 1991, 1992, 1994, 1997, 1998; Labov & Fanshel, 1977; Wodak, 1981) point the way for an investigation of the dynamic interaction underlying discourse, the co-construction of discourse (Duranti & Brenneis, 1986). Researchers (e.g., Langs, 1981) are beginning to view resistance as a cooperative failure between client and therapist goals with contributions from both therapist and client. Most previous empirical work has not examined the contribution that therapists make to client resistance. In this chapter those dynamics are probed through discourse analysis of 26 hours of tape-recorded individual psychotherapy sessions between five clients and four therapists (both male and female) along with ethnographic observations including feedback sessions, attendance at training sessions, viewing through one-way mirrors, and interviews with both clients and therapists. Each client was recorded in six consecutive sessions occurring once a week over the course of 6 to 8 weeks. The focus of this chapter is on repeated verbal expressions of resistance by clients to therapeutic suggestions by therapists such as to *let yourself cry*. The chapter traces how these and other therapeutic suggestions are met linguistically. Certainly not all resistance is observable. Covert or strategic resistance may not be measurable. But tactical resistance has outward signs and is immediate and observable (Mahalik, 1994). This chapter examines the outward verbal expression of resistance. An assumption is that if we can pinpoint how and where resistance is occurring, make its linguistic manifestation more precise, recognizable, and measurable, then a skilled psychotherapist or trainer may be able to develop countermeasures to reduce resistance and to free up the expression of emotion.

Health care providers of all sorts frequently make therapeutic suggestions. In a medical setting, for examples, physicians may suggest that a patient take medicine or follow treatment plans such as taking pills regularly, getting exercise, or changing diet. In the psychotherapeutic setting, therapists may suggest activities to enhance insight or facilitate change in line with agreed upon goals. However, both physical and mental health care providers report that often their patients and clients are noncompliant. Noncompliance or client resistance is a central concept in models of psychotherapy ranging from analytic psychotherapy to cognitive-behavioral, gestalt, and rational-emotive psychotherapy. As Kolko and Milan (1983) reported, there is a continuum, with the actively oppositional client and the actively compliant client on the two extreme ends and the inert client in the middle.

Therapists want to minimize resistance but have barely begun to operationalize what it is. Often in psychology studies, after coding and tabulating of therapist and client response modes, the actual language data disappears after holistic judgments are rendered. In this chapter I take a discourse-centered approach using actual language samples recorded in individual psychotherapy sessions to begin to pinpoint how and where resistance is occurring, and what, if any, steps a skilled psychotherapist might take to reduce it. For the goal of this chapter, to examine how client resistance manifests itself linguistically, it will be helpful to bear in mind that researchers such as Langs (1981) see resistance as a cooperative failure between client and therapist goals, with contributions from both client and therapist in its creation and maintenance. We later examine evidence that supports this.

DEFINITION OF RESISTANCE

According to Schuller, Crits-Chistoph, and Connoly (1991) and Mahalik (1994), resistance is a multidimensional phenomenon. Several dimensions are intertwined, all in the service of avoidance of painful feelings. Resistance is present when clients block expressions of painful feelings and withhold personally revealing material. This can also involve communicating verbally but with an absence of or minimization of affect, even when events that would be expected to be highly charged with emotions are involved. Clients can also oppose recollection of material because it is painful. The client may provide vague versus detailed information to the therapist about himself, keeping conflicted material out of the conversation and failing to share relevant information. Instances of silence, saying *I don't know*, or saying he or she doesn't feel like talking, are frequent. Typically clients withhold personally revealing information while seeming to communicate. This can be accomplished by talk about inconsequential events, using clichés, or glossing over events without giving details.

Another dimension we examine shortly is opposing the therapist when the therapist proposes some therapeutic task. A client's resistance may take the shape of a contrary response to the therapist's requests or interventions. The client may attempt to undermine the treatment efforts by the therapist or oppose the therapist's attempts to change the client's patterns of thinking, feeling, and acting, thus maintaining the problem, according to Mahalik (1994). Likewise, the clients can oppose change. Even though clients have sought help and consented to cooperate with the therapist, they may show evidence of resistance in opposing change and express a wish to maintain the status quo. They hinder or interfere with change in spite of indications given by them that the current situation is maladaptive, limited, or painful. Similarly, they may oppose insight. Insight is the making of connections between experiences, thoughts, feelings, and behaviors that leads to self-understanding, a goal of all therapy. The motivation for avoiding insight, according to Mahalik, is to avoid the painful feelings that accompany sometimes disturbing insight. Several illustrations of these dimensions are provided with examples from the corpus.

BACKGROUND TO ANALYSIS

In order to explicate the face of resistance I draw on several key notions developed by discourse analysts. Of particular import are the notions of *repeated requests* and *put offs* developed by Labov and Fanshel (1977). The next section examines several cases from five different dyads. Particular attention is paid to one striking instance that illustrates nine put offs of a repeated request in an 8 minute period. Work by Weijts (1993) shows the risk caused by use of indirect requests in a gynecological setting. Weijts showed that patients' indirect requests were often ignored by physicians. I concentrate here on the risks to the interaction as a result of therapists' use of indirection. Efforts to use deference and mitigation to equalize the power imbalance and build rapport can have deleterious effects when clients seize the therapist's indirection as a ploy in their resistance.

In order to achieve their goal of increasing client self-awareness and effecting change, trained and licensed psychotherapists (usually clinical psychologists with doctoral degrees, also psychiatric social workers) occasionally make therapeutic suggestions to clients about behavior they feel would be in the client's best interest. These are well-considered requests gauged to be therapeutic based on the therapist's extensive training and experience as a mental health professional. As such, it is valuable to recognize that these therapeutic suggestions bear similarities to everyday advice-giving which may only be surface similarities. Unlike the work of radio talk show therapists discussed in Gaik (1992), these are not instances of advice or counseling about life. These are suggestions that are linked directly to therapeutic effects. Neither should we consider these therapeutic suggestions the same as spontaneous lay advice-giving (see Jefferson & Lee, 1981, for a cogent analysis of lay advice-giving in response to troubles-telling).

EXAMPLES OF THERAPEUTIC SUGGESTIONS

Some examples (in 1 a-e) will illustrate the range of therapeutic suggestions found in the corpus. For example in 1a the therapist says, at the arrow, *I'd like to suggest you just let yourself cry.* Certainly a goal of therapy is free expression of affect and therapists work diligently to foster this freedom of expression. However, in 1a the client is resistant. She doesn't want to express herself. She holds back her sobs when requested to express her pain.

(1a) GW(14)[1]
(Client is telling story about how her mother used to abuse her father. She is fighting back sobs.)
[T = therapist C = client]
→   T: I'd like to suggest you just let yourself cry.
      C: Huh? ((crying))
→   T: I think it'd be a good idea if you just let yourself cry.
      C: I thought I did real well at that ((laugh))

---

[1]Examples have been transcribed using the system described in Ferrara (1994, Appendix B). Short unfilled pauses are marked by "..."; longer pauses have the delay times in seconds in parentheses. Nonverbal behaviors such as laughs and crying are indicated in double parentheses. Underlining is used to indicate emphasis. Elongated syllables are marked with a colon. Latched speaking turns, in which the second party starts to talk with little or no delay after the completion of the first party's turn, are marked with "=." Brackets indicate overlapping utterances by two separate speakers. See Ferrara (1994) for a more detailed description of transcription conventions.

T:	Huh?
C:	I thought I did real well at that ((laugh))
T:	Well (.) I don't know what you may do outside of here but it looks like you're really fighting yourself right now.
C:	(3) I don't like to cry. I feel like that's all I been doing is crying ((sniff)). I fight it back (.) sometimes ((sniff)). I race home and cry at home.

The example shown in 1b is a very effective series of requests. Notice at the five arrows the therapist does not use indirection in his requests. He asks the female client to do an exercise in which she is to rate the types of relationships she has. Notice the forms the therapist uses for his requests. They involve statements using *I'd like you to* and *I want you to*. (In a later section I discuss Ervin-Tripp's (1976, 1984) claim that this is an effective request structure.) In this example the client agrees with the therapeutic suggestion with a series of *Okays*. She does comply by bringing in a list of six relationships in the next recorded session. She is both verbally and behaviorally compliant.

(1b) GW2(38)
(Client is discussing two warring sides of herself)

	T:	There's something I'd like you to do:
→		
	C:	Okay.
→	T:	I want you to do some homework. Uh
	C:	Okay.
→	T:	I want you to (4) take uh ten important relationships to you
	C:	Take ten important relationships?
→	T:	Yeah I want you to take, like Martin, two children, separate relationships you and your brothers and sisters had (.)
	C:	My mother, my father, whatever
	T:	Uh maybe some people at work
	C:	Okay
→	T:	Sister, and I want you to uh like in one column put their name, and in the second column I want you to put which part of you is most in (.) uh (3) most there in that relationships. Is it the cold aloof part or is it or is it the soft warm-hearted?
	C:	So that's that's all you want me to put as to whether it was the soft me or the-
	T:	Yeah.
	C:	Okay.

In example 1c much the same success occurs. Here the therapist makes a therapeutic suggestion for the client to seek information about her boyfriend's reaction to her weight. The therapist also couches these requests in the terms *I'd like you to*. Again, the client verbally agrees with *Okay* and in fact does comply the next week.

(1c) GW6(16)

→   T: ...Well there's something I'd like you to ask Martin.

    C: Okay.

→   T: And I'd like you to ask it verbatim (1) you know like like I'm saying. In fact I think I'm gonna write it down so I can have you ask it in the exact words.

    C: Okay. So I don't forget it.

    T: Huh?

    C: Just so I don't forget it.

→   T: Well, yeah 'cause the words, sometimes a single word can make a difference. And I'd like you to look him straight in the eye and say, "Martin, you told me that sometimes you want me to be fat. (2) What would be too fat?" (2) And I'd like you to have him describe what "too fat" would be...in great detail what too fat would be.

    C: Okay.

However, in example 1d the same success does not apply. Here, a different therapist uses a mixture of forms to make a request, the majority of which are worded indirectly. He tries to get his male client to decouple smoking and drinking. His client has stated that his goal is to quit drinking and the therapist feels that the smoking leads to the excessive drinking. In attempting to de-link the two habits, the therapist suggests replacing the urge to smoke with a session at the computer. His indirect question *Do you think you'd be able to uh get on your computer when you uh have an urge to drink and smoke?* elicits a rather noncommittal *I suppose I could, yeah.* Here we are concerned with the verbal rather than physical (non) compliance. Even after 7 weeks, the therapist does not gain compliance with his requests. More importantly, there is verbal foreshadowing of this. Notice the client never says *Okay*, and he uses evasive language such as *I suppose I could*, and does not clearly verbally commit to the action.

(1d) RK1

→   T: That's why I think it's very important in *your* case to discontinue smoking *and* drinking.

    C: Yeah.

→   T: And probably uh it might be more helpful for you to permanently stop smoking and then it might help you to avoid some of these slips that you have in the drinking.

    C: Yeah.

...

(discussing "positive addition," substituting something constructive for something destructive)

→   T: You think you'd be able to uh get on your computer when you uh have an urge to drink or smoke?

     C: I suppose I could, yeah.

...

→   T: Okay. Well uh I'd like you to try it using your computer as the substitute for smoking first. Try that for a week or two (.) a:nd if you find that you're - First of all let me just remind you that (.) the research that's talked about in there says that most withdrawal symptoms go away (.) in about one to two weeks.

...

→   T: Do you want to try to go for a couple of weeks without smoking and use your computer?

     C: Sure.

     T: ((writing)) (2) use computer when urge to smoke occurs. I'll put two weeks.

(7 weeks later)

RK(8)

     T: Did you try drinking without smoking?

     C: No, I haven't.

     T: *Not* yet ((laugh)). Okay.

     C: I'm not sure what it is cause somehow I just can't seem to separate the two in my mind.

Example 1e is a lengthy one in which the therapist makes a therapeutic suggestion to his female client who had been raped in childhood that she report a recent rape attempt on her by an acquaintance. He feels that expressing now what she could not express then as a child would be therapeutic. The outrage, terror, and shame felt by an inchoate 10-year-old can finally find a voice in the same emotions of his adult client, he believes. The client initially expresses resistance to the idea, *But I don't want to deal with that ... I don't want to press charges ... I don't want to be involved with that stuff.* The therapist uses a series of suggestions with *would, could, might,* some worded indirectly, some more directly.

(1e) RL(2)

(discussing a recent attempted rape of a woman who had also been raped in childhood)

	C:	I wonder if I should file a report=
→	T:	=exactly what I'm asking right now. For your protection, for the protection of other women.
	C:	But I don't want to deal with that. I don't want to press charges. All all I would want to do is file a report because
	T:	Mmhmm
	C:	I don't *want* to be involved with that stuff.
	T:	I can see that. I can see that.
	C:	((throat clear))
→	T:	Uh (.) I'd suggest that you seriously consider doing that. Because it sounds like he is a potential rapist. And may actually [have raped.]
	C:	[((throat clear))] Yeah. He's *been* in trouble [before.]
	T:	[Yeah] Yes.
	C:	He's been in trouble before.
→	T:	I think the descriptions that you could give of this individual um would be very helpful (.) to law enforcement authorities. To help (.) women in a similar place.
	C:	Yeah. I know where he lives right now and I know his first name.
→	T:	Well I think that would be (.) important. Uh You might also check to see
	C:	And I bet. I think they're aware of him. He goes- is supposed to go to MHMR once a month and they told him he needed to go to group and this isn't ( ) group. And it's kinda like, it really might help me a lot if I go through with it but I wasn't gonna evaluate anything he had to tell me. I was just gonna go, "oh."
→	T:	If it would make you feel comfortable too, you might ask one of the men (.) that confronted him to also (.) file a report with you. On the other hand you could take responsibility for it yourself
	C:	Uh huh
	T:	depending on what feels good to you. Since I think the AA program, my view of it is that is to confront people=
	C:	=with their behavior=
	T:	=in a healthy way, this their behavior. And too in that sense, really help people to (.) help themselves.
	C:	Mmhmm.

T: So it might be something that the AA club itself could (.) do jointly.

C: Oh I you know I think they'd tell- there's still I think in AA there's a difference between taking care of your club and the people in it and going to cops.

T: I can see that.

C: There's a difference (.) in there.

T: I guess the part of me though I guess speaking to you (.) as your counselor is that I'm concerned for you and for [other women] in that situation.

C: [other people]

T: And I think that goes beyond (1) how shall I say, protecting confidentiality. He broke the law.

C: And I'll just file a report and tell them as involved as I want to be.

→ T: Why don't you call anonymously and ask to speak with yeah (.) say that you'd like to (.) file a report anonymously. Just ask under what circumstances you can do this. I think that would probably fall under the category of one of those – what do they call it - crime stoppers (.) type of. You might have seen them in the newspaper.

C: Oh. Right. Yeah.

→ T: Where you call without giving your name. At least you can describe the individual. You can describe what happened. And then=

C: =I think they have him on file.

→ T: And if you would like (.) to call me and tell me what (.) what the result was that will just alleviate yourself (.) of the uh (shield) of responsibility to uh (.) protect yourself. Because we have far too many rapists around these days.

C: Mmm under pressure.

→ T: You might also call and mention MHMR. An uh indicate (.) that's (.) describe him, the first name and the circumstances and inform them that if any counselor knows of a gentleman that meets that description in a group that this was the situation that happened.

C: And it happened to another woman.

T: Yes.

C: Uh That I was *told* it happened to another woman.

T: The same gentleman did this.

C: Yeah.

T: That's pretty serious. (1) Well I hope [that you'll understand my] making this kind of

C: [I had - After I got through the] trauma of that shit uh a
woman at work come up to me. And that was my next issue.
That was my next pending issue.

T: Which is?

C: What to do with that. [Whether] to call the police or not.

T: [what to]

C: So I'm glad you brought that up.

→   T: *I* think it's important. I also think that you have had a
valuable experience from this (.) reliving your earlier abuse
when you could *not* talk about it.

These extracts are offered as background context to indicate the type of
therapeutic suggestions that occur and some of the ways in which they are
worded and reacted to.

KEY CONCEPTS FROM DISCOURSE ANALYSIS

We turn now to examine the underlying concepts that will illuminate a rather
dramatic sequence of putting off requests. In their pioneering study of
Therapeutic Discourse, Labov and Fanshel (1977) posited that indirect requests
were based on questions or assertions about

(2) Various Conditions Underlying Requests

    a.   the Existential status of an action by B
    b.   the Consequences of such an action
    c.   The Time the action might be performed

Or d.   The preconditions
           Need for the action
           Need for the request
           Ability to do X
           Obligation to do X
           Willingness to do X
           Right of A to ask B to do X

These insights about the relationship of requests to indirect requests are
shown in (3) and (4). They express generalities or regularities in terms of rules.

(3) RULE OF REQUESTS

If A addresses to B an imperative specifying an action X at Time T1 and B
believes that

1a.	X should be done (for a purpose Y) *need for the action*
1b.	B would not do X in the absence of the request *need for the request*
2.	B has the ability to do X (with an instrument Z)
3.	B has the *obligation* to do X or is *willing* to do it
4.	A has the right to tell B to do X,

Then A is heard as making a valid request for action.

(4) RULE OF INDIRECT REQUESTS

If A makes to B a Request for Information or an assertion to B about

1.	the existential status of an action X to be performed by B
2.	the consequences of performing an action X
3.	the time T1 that an action X might be performed by B
4.	any of the preconditions for a valid request for X as given in the RULE OF REQUESTS

and all other preconditions are in effect, then A is heard as making a valid request of B for the action X.

Discourse analysts have found Labov and Fanshel's (1977) rules to be extremely valuable. Even more germane for our inquiry into resistance is the Rule for putting off requests that Labor and Fanshel explicated. They claimed that interlocutors can successfully refuse a request by asking a question or making a negative assertion about such things as Need, Time, Ability, Obligation, Willingness, and Right to Ask. Their generalization is shown in (5).

(5) RULE FOR PUTTING OFF REQUESTS

If A has made a valid request for the action X of B and B addresses to A

1.	a positive assertion or request for information about the existential status of X
2.	a request for information or negative assertion about the time T1
3.	a request for information or negative assertion about one of the four preconditions in RULES OF REQUESTS (need for action, need for request, ability, obligation, willingness, right to ask)

then B is heard as refusing the request until the information is supplied or the negative assertion is contradicted.

The regularities captured in (5) will form a basis for our analysis of the Main Text. The Rule for putting off requests shown in (5) provides a possibility of nine different choices for ways a person might put off a request. We can also add two other resources from Grimshaw (1982). He posits in (6) that listeners may use the strategies of Spurious Misunderstanding and Spurious Non-Hearing to sidestep requests. That is, hearers can hear and understand but *act as if* they had not. They may *act* too dense or too literal minded and if caught can apologize or justify.

(6) Grimshaw's (1982) notions of SPURIOUS MISUNDERSTANDING and SPURIOUS NON-HEARING
Where hearers hear and understand but *act as if* they had not.

a. Spurious Non-hearing, according to Grimshaw (1982) is viable as an H response to S communicative attempts because there are instances in which understanding is not experienced because maxims are flouted, indirection is too subtle, or H is too literally minded.

b. Spurious Misunderstanding can vary on 3 continua: (1) overtness-covertness (they may be blatant and obviously intended, recoverable but ambiguous or well-concealed); (2) friendliness-unfriendliness (they can be whimsical, teasing or punitive, self-protective, or outright malicious); (3) excusability-justifiability (if caught the person can apologize or justify).

The generalities captured by Labov and Fanshel (1977) and Grimshaw (1982) show how a request may be put off, but no one has predicted that anyone could use all these strategies at once! We see later that giving an accounted refusal to an indirect request leaves room for a repeated request, a recycling or recursion.

BACKGROUND TO ANALYSIS OF MAIN TEXT

Let's turn now to the Main Text to view a rather dramatic sequence of putting off requests as a type of resistance. These put offs are accomplished by one client in about 8 minutes. This psychotherapy client, pseudonymed Jeff, is an engineer in his late thirties who works in computer sales. He has recently separated from his abusive wife. He is seeing the therapist-in-training Bonnie, an advanced doctoral student in clinical psychology. She is supervised by a licensed psychologist and is working in a clinic, where Jeff has sought professional help. Six consecutive weekly sessions were recorded and transcribed. An appendix at the end of this chapter provides relevant background segments from earlier in the same therapy session to show that both client and therapist have together established that the client is overworking and needs to relax. A look at the background segments reveals that the client admits that pressure is building up (A), one day he'll just have to sit down and relax (B), he deliberately keeps busy to avoid thinking (C), he's overly conscientious and uses overwork to stay busy and avoid facing his problems (D & E), and the weight of the world is on him (F).

In the Main Text we focus on the role of indirection in facilitating a recursive and unproductive cycle of putting off requests. The text illustrates the face of resistance. The discourse abstracts shown in (7) are taken from one 50-

minute session between the client Jeff and the therapist Bonnie. There are nine highlighted segments, all consecutively occurring in one session. These segments flow into each other, sometimes with extraneous material edited out. For convenience the segments have been numbered with Roman numerals. In them the therapist gives a series of indirect requests and the client skillfully resists, putting off each one. Our intent is to highlight the discourse dimensions of the co-construction of resistance and to examine the verbal resources used by the therapist and client in tandem. Throughout the analysis of these nine segments I appeal to the work of Labov and Fanshel (1977) and Grimshaw (1982).

In the Main Text, the crux of the interchange is that, following up their previous discussions earlier in the session, Bonnie is requesting that the client Jeff find a way to relax and not work so obsessively. She believes that if he relaxes he will then be able to process the many conflicted emotions he is experiencing in connection with a pending divorce. She believes that being a workaholic is a ploy to avoid processing painful emotions. This belief is echoed by Miller (cited in Restak 1988) that addictive behavior of all sorts, "be it cocaine, heroin, alcohol, nicotine, gambling, sexual addiction, food addiction, all have one common thread. That is the covering up or the masking, or the unwillingness on the part of the human being to confront and be with his or her human feelings" (p. 106).

Consider the nine segments in the Main Text shown in (7).

(7) MAIN TEXT—PUTTING OFF REQUESTS

I		
1	C:	I like to play pool. It's relaxing. Uh (.3) Get together with friends have
2		cook-outs, you know
3	T:	Mmhmm
4	C:	Water-ski. I got me two friends that have boats.
5	T:	So you have things you *can* relax with?
6	C:	Mmhmm. It's just a matter of putting the work aside (.3) and going and
7		doing the relaxing part.
II		
8	T:	What's keeping you from doing that right now?
9	C:	I wouldn't. I think that (.3) uh if I work hard enough I'll just, you know,
10		forget about it.
11	T:	Is that=
12	C:	=maybe I'm trying to punish myself, say, well, you know,

| 13 | | "You don't need to relax." You know, you just- I don't know. It's a little confusing |

III

14	T:	When do you think you'll come to the point where you can deal with this?
15		Instead of burying yourself in your work so you won't have to think about it? ((soft voice))
16	C:	Uh (2.0) probabl:y, like I saw [after it's all settled up]
17	T:	[after you've settled]
18	C:	Yeah. Cause right now it's too much up in the air. You don't know=
19	T:	=Mmhmm
20	C:	=what's what.

IV

21	T:	Would you be able to take off, though, even if this settlement is not? Could
22		You take off before you married her, when you didn't, when you weren't steady
23		Dating with her?
24	C:	Could I take off and you mean =
25	T:	= and relax and not worry about your job?
26	C:	Uh (3.0) Yeah. (1.0) Uh, you know. Yeah. She had a lot of problems with one
27		Job when I first met her, so, working for an employment service and it was (.3)
28		Very political. Everybody was very push. And you would go in maybe to pick her
29		up for lunch or something and you could, you could just feel it when you walked
30		in. You know, it was like there was knives in your back or something. And I had
31		some friends that went in ( ) dropping something off for her, for me,
32		cause I couldn't do it. And (.3) when they walked in like, "Boy
33		I don't want to go back in there."
34	T:	Yeah ((CC continues digression, talking about wife's eye surgery problems)) (portion omitted0

V.

35	T:	Could you *ever* get away from the problems though. I mean *could* you relax?
36	C:	Uh (.5) at times, yeah. (4.) But it go worse and worse, you know.
37	T:	Mmhmm.

38	C:	And (1.) as soon as you'd walk in the door it was harp, harp, harp. You know (.5)
39		from the time you got there. (1.) You'd walk in, and you know, the first thing
40		she did, she was waiting to gripe. And she'd expect me to fix supper.

VI

41	T:	Do you think there is any way you can relax somehow ((laugh)) this weekend?
42		Can you take *any* time off=
43	C:	=Unh-uh=
44	T:	=to water-ski or play pool?
45	C:	No, not pool. ( )

VII

46	T:	Not even an hour?
47	C:	Uh (.5) I think probably (.3) I'll just call one of my friends (.3) and go out
48		and just do something.

VIII

49	T:	I think that would probably be really good to do that at least *once* this week.
50		Try to get some of that off your shoulders=
51	C:	=Yeah=
52	T:	=Just relax for a while.
53	C:	yeah, I'd *like* to do that, that's (.5) that- I just keep thinking, one of these
54		days.

IX

55	T:	Try to make it this *week*, even if it is just an hour ((laugh)). It will probably
56		make you feel better
57	C:	Well, what it if doesn't?

ANALYSIS OF MAIN TEXT

Of course, explanations for *why* a client expresses resistance are left open to another field. Our interest in this analysis is in *how* resistance manifests itself. In Segment I, the therapist questions Means and Ability. Following the client's claims (lines 1-2) that he plays pool and enjoys outdoor activities such as boating, water skiing, and cooking out, the therapist addresses a question with declarative sentence pattern and final rising intonation (line 5), *So you have things you can relax with?*   Her question falls within the category of Ability

(Ability to do X with instrument Z) and is directed at the Means the client has available to him to relax with. The therapist's remark begins with *So*, which entails that both therapist and client have mutually acknowledged (Ferrara, 1997) that the client has the Means to relax. In his next utterance, *It's just a matter of putting the work aside (.3) and going and doing the relaxing part* (lines 6-7), he refers to the precondition of Willingness. He says, in essence, it's not that he can't, it's just that he hasn't wanted to enough to make the effort. The effortful part, he states, involves *putting the work aside* (line 6) and *going and doing*. The client makes the relaxing part sound like another task—effortful.

In Segment II (continuous to Segment I in the recording) the therapist questions Ability and Willingness at a Specific Time. The therapist's *What* question, *What's keeping you from doing that right now* (line 8) is a direct response to the client's previous remark involving Willingness. To ask what prevents someone from action could elicit a variety of responses. The therapist has stipulated the immediate time frame by her phrase *right now*. Her question (line 8) implies that she believes the action (going and relaxing with the aforementioned recreational possibilities) would be beneficial, which implies that the action should be done soon.

In responding (lines 9-10), the client states that he may *just forget about* his Need or Desire to relax. If he *works hard enough* (line 9) the need may go away. The client indicates that the wish to relax is a temporary temptation, not something part of an ongoing strategy.

The therapist attempts to restate her initial question (from line 8), *Is that=* (possibly to be completed with *what's keeping you from doing that right now?*), but she is stopped short after two words by the client's latched insightful self-talk. Out loud he muses, *Maybe I'm trying to punish myself* (line 12), say (to himself) *well, you know, 'You don't need to relax'* (lines 12-13). The client's comment *I don't know* is also a classic resistance line, according to Mahalik (1994).

The issues in the segment are especially complex and no doubt there is more going on than a response to the therapist's question. It appears that the client is having a dialogue with himself, out loud, over some internal struggle. We can surmise that perhaps part of him wants to relax and part of him demands that he work. Nonetheless, when some part of his self tells the other *You don't need to relax* (line 13), a doubt about Need is expressed. This too is one of the preconditions Labov and Fanshel (1977) predicted to be an accounted refusal.

In Segment III (some 46 lines separated from Segments I and II in the recording), the therapist questions Time with a remark containing *When?, When do you think you'll come to the point where you can deal with this?* (line 14). According to Labov and Fanshel (1977), this is a request for Information about Time1 that an action might be performed by B. In the latter part of line 14 and line 15, the therapist again repeats the central issue underlying the repeated request—the fact that the client hides behind his work. Her use of the phrase *burying yourself in your work* (line 15) implies that the client is not fully living, thereby challenging the client's role as a healthy person. The therapist softens this by modulating her voice.

The client reacts to this implied challenge with a hesitation of two seconds (line 16), *Uh (2.0)*. His reference to an indefinite future Time, *after we've settled up* (line 16) is a polite refusal, which contains an accounting. He puts off the therapist's indirect request by an appropriate comment about future (unspecified) Time. His statement implies that he will comply with the request after a further precondition is met, namely, after he has completed the arrangements for a financial settlement with his wife, which is currently *up in the air* (line 18). Important to keep in mind here is the tenet Labov and Fanshel (1977) stated as a fundamental mechanism of discourse: "In denying or questioning one of the preconditions for the request, B implicitly accepts all of those preconditions that he has not denied or questioned" (p. 88). By stating that the Time is not right (lines 16 and 18) at the moment to comply with the therapist's request, the client tacitly accepts the premise that the request is a valid one.

Apparently not satisfied with the accounted denial hinging on Time (when the settlement is completed), given in Segment III, the therapist re-addresses her request in Segment IV, expressly questioning Ability Regardless of Time. In line 21 the therapist discounts the Time excuse (*even if this settlement is not*, meaning "is not finished") and phrases her request in terms of Ability: *Would you be able to take off, though, even if this settlement is not?* (line 21). She follows this request with a rephrasing in similar terms of Ability, *Could you take off?* (lines 21-22) again, regardless of Time, *before you married her* (line 22). The undertone of her persistent question is indicated by the use of both the future and past tenses.

The insistence of the therapist's follow-up, with two questions of Ability Regardless of Time juxtaposed, can leave little doubt of the point she is making, yet the client (line 24) appears to be trying to "buy time" in rephrasing her question, *Could I take off and you mean*. She includes the reason for the main thrust of her intervention in the two conjoined verb phrases she latches on, *and relax and not worry about your job* (line 25).

In Labov and Fanshel's (1977) terms, following Goffman, this exchange is face-threatening because the therapist has, in essence, rejected the accounted refusal the client offered in Segment II, because she has set Time aside as a valid excuse.

The client's response may be accounted for using Grimshaw's (1982) notion of Spurious Non-Hearing. The client has certainly heard the double-edged question but chooses to ignore the therapist's challenge. Instead, he bridges over into a personal narrative that does not seem pertinent. The client allows himself to ruminate (lines 26-33) on a past situation his wife was involved in at a time prior to their marriage, *when I first met her* (line 2). His tactic, in the face of a challenge, is to sidestep the issue at hand by recounting a tale about his wife's work tensions, thereby attempting to refocus the conversation away from himself. This refocusing away from self is another resistance tactic, according to Mahalik (1994).

Segment V follows the exchange in Segment IV with only 14 lines intervening. For 14 lines the therapist allows the client to digress about his wife's early experiences. In line 35, however, the therapist attempts to turn the conversation back to the same question she posed in lines 21 and 22, the question of Ability. She repeats the same modal expression used in line 21, twice more, *Could you ever get away from the problem, though, I mean, could you relax?* By not varying the form of her request, insistent *could you's*, she indicates she does not want to be put off and that the client should function properly in his role as a client. The therapist uses the attention-catching device, *I mean,* to reiterate her question of Ability, *I mean could you relax?* (line 35). The client must respond to such a challenge. He can no longer pretend not to have heard. Instead he hesitates (line 36) and replies with a comment about Ability at Indefinite Times in the past *at times, yeah,* along with a 4-second pause. Following the awkward pause, he changes topics to avoid further discussion of his abilities. He launches instead into another attempt to digress about his wife and her nagging (lines 38-40).

The succeeding four segments (Segments IV-IX) occur in the final 3 minutes of the 50-minute therapy session. The therapist evidently feels some urgency about repeating her request. She is sincere in her belief that the therapeutic suggestion for action will benefit the client and she continues to press the issue.

In Segment VI the therapist uses another categorical expression, similar to her use of *ever* in line 35. She repeats the word "any" in lines 41 and 42: *any way* and *any time.* In line 41 she questions Ability at a Specific Time in the near future, *this week. Do you think there is any way you can relax somehow this week?* In this query she leaves the Means indeterminate, saying *somehow,* but in the lines immediately following (lines 42 and 44) she no doubt recalls the favored types of recreations previously mentioned by the client in lines 1 and 2 of Segment I. She appends two specific Means to her main indirect request regarding Time and Ability: *Can you [Ability] take any time [Time] off,* [latched with Means], *to water ski or play pool?* The client does not respond to the Time or Ability portions of her indirect request. Instead, he uses what Grimshaw (1982) calls Spurious Misunderstanding. The question as posed in lines 41 and 42 was certainly not one regarding which of two alternative modes the client might prefer, but the client utilizes the technique of literal interpretation of the last two words in the therapist's indirect request. The wily client *acts as if* the issue at stake is whether or not he wants to play pool and ignores the main point of the therapist's question when he replies, *no, not pool* (line 45).

It is interesting to note that psycholinguistic research, according to Ervin-Tripp (1984) and Gibbs (1979), concludes that it is unlikely that interpretation of indirect requests includes a literal phase. Both Gibbs and Ervin-Tripp argued that it is not necessary to first compute the literal meaning of an indirect request.

This particular rejection of an indirect request under the guise of literal interpretation or Spurious Misunderstanding is not a well-concealed one and the therapist immediately counters in Segment VII (which is continuous and contiguous with Segment IV) with a challenge about the Time demands, implying that anyone could spare an hour for his own benefit (line 46): *Not even an hour?* This translates as, *No? Do you mean to tell me that you can't even escape for a mere 60 minutes?*

Unable to refuse such a reasonable request outright, the client takes refuge in the use of indefinite references. These are noncommittal. He uses phrases *probably, one of my friends, do something.* He makes no definite plans, gives no ground. Still after seven tries by the therapist the client has not agreed to the therapeutic suggestion. This is an accomplished put off. The client shows himself to be a resourceful speaker.

In Segment VIII the therapist refers to Consequences (presumed benefit) and Time. She comments, *That would probably be really good* (line 49), to follow the recommended action at a specific Time and frequency (*at least once this week*). For the first time, line 50, she gives an indirect command to the client. Note the imperative if *Try to* is removed:

> Try to get some of that off your shoulders (line 50)
> Get some of that off your shoulders

She follows up with a more direct command, *Just relax for a while* (line 52). This departure here from the previous exclusive use of indirect requests underscores the therapist's belief that the client should comply with her request. Having failed with indirect requests, the therapist is now willing to use her authority (as accepted by both the interactants) to give an imperative about action in the client's best interest. As predicted by Labov and Fanshel (1977), one further means of putting off a request is to refer to Willingness. In line 53 the client, having heard the imperative, responds, *Yeah, I'd like to do that ...* The implication after *like to* is *but...* meaning but something else prevents me besides my willingness. In lines 53 and 54, the client refers to an indefinite Time in the future when he will comply with the request. His wording, *one of these days*, is just as noncommittal as was his phrasing in line 47, *one of my friends*. Recall that in Background Segment B (line 14) the same wording, *one of these days*, was also used to temporize.

In the final segment, Segment IX, with the allotted session time up, the therapist again issues an indirect command with *Try to*, (line 55). She re-refers, as in Segment VIII, to the specific Time, *this week*, and limited duration, *just an hour* and adds a comment in the next sentence about the presumed beneficial

Consequence (again, as in XIII): *It'll probably make you feel better* (lines 55-56).

Instead of agreeing, *Okay, I will*, as one might expect of the client at the conclusion of the therapy hour and a sequence of eight earlier put offs, he again puts off the request by addressing a question to the therapist about the Consequences of the action. Startlingly, he responds, *Well, what if it doesn't?* (line 57). Unbelievably, I submit, he has once more succeeded in putting off the request. This is his ninth successful put off in fewer than eight minutes. We have to grant him skillful use of linguistic resources to achieve put off. He successfully resists the therapist and uses a full repertoire of accounted refusals.

In summary, we have viewed a recursive sequence of indirect requests and put offs. The predictions of Labov and Fanshel (1977) and Grimshaw (1982) were seen to have a high degree of generalizability to a novel situation. As predicted, reference was made in the refusals to Need for Action, Ability to do the action, Means, Willingness, Time, and Consequences (benefit). The linguistic formulas described by Labov and Fanshel were exemplified in the nine segments displayed. In addition, Grimshaw's notions of Spurious Non-Hearing and Spurious Misunderstanding were shown to have genuine plausibility in accounting for those areas not directly explainable by Labov and Fanshel's rules.

Figure 12.1 presents this recursive chain of requests and put offs in a schematized form. In summary, we can see in Figure 12.2 a series of indirect requests by the therapist has led to a recursive chain of put offs and repeated requests. The schema highlights the linguistic resources used by the client. The mutual responsibility of both therapist and client in co-constructing this recursive chain is evident in Figures 12.1 and 12.2. Each step by the other results in yet more unsuccessful maneuvers by the other.

An alternate but not necessarily contradictory explanation for the surprising sequence jointly engaged in is provided by the findings of West (1983, 1984). Her research points to two constraining structural circumstances in the ways that doctors formulate their questions, which thereby limit patients' options for answers. These are chained, multiple choice answers and use of either/or questions. In both cases, the respondents are not clear to which, if any of the questions they should venture a reply.

Observe how the very design of several of the instances given leaves little possibility for the client to respond. In three segments, Segments IV, V, and VI, the therapist asks two questions latched together: lines 21 to 23 *Would you ... Could you ..*; line 35 *Could you ... Could you ..;* lines 41 to 42 *Do you ... Can you ...* This tactic may leave the client puzzled as to which question to address, or at least able to pretend that he is puzzled. Likewise, in Segment VI, the latched forced choice alternative structure, line 44, *to water ski or play pool,* may contribute to the ineffectiveness of this sequence.

Therapist   Client

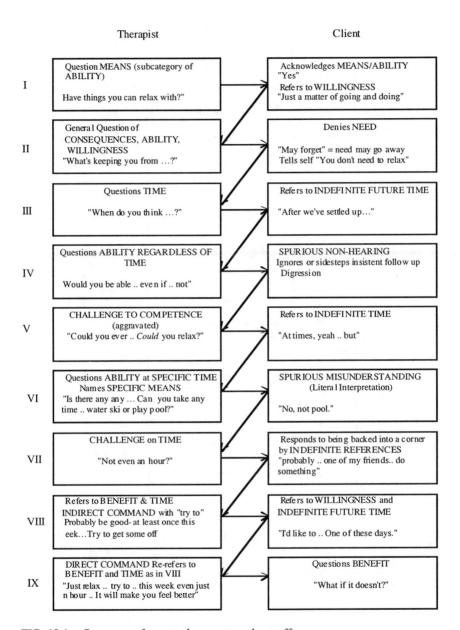

	Therapist	Client
I	Question MEANS (subcategory of ABILITY) Have things you can relax with?"	Acknowledges MEANS/ABILITY "Yes" Refers to WILLINGNESS "Just a matter of going and doing"
II	General Question of CONSEQUENCES, ABILITY, WILLINGNESS "What's keeping you from ...?"	Denies NEED "May forget" = need may go away Tells self "You don't need to relax"
III	Questions TIME "When do you think ...?"	Refers to INDEFINITE FUTURE TIME "After we've settled up..."
IV	Questions ABILITY REGARDLESS OF TIME Would you be able .. even if .. not"	SPURIOUS NON-HEARING Ignores or sidesteps insistent follow up Digression
V	CHALLENGE TO COMPETENCE (aggravated) "Could you ever .. Could you relax?"	Refers to INDEFINITE TIME "At times, yeah .. but"
VI	Questions ABILITY at SPECIFIC TIME Names SPECIFIC MEANS "Is there any any ... Can you take any time .. water ski or play pool?"	SPURIOUS MISUNDERSTANDING (Literal Interpretation) "No, not pool."
VII	CHALLENGE on TIME "Not even an hour?"	Responds to being backed into a corner by INDEFINITE REFERENCES "probably .. one of my friends.. do something"
VIII	Refers to BENEFIT & TIME INDIRECT COMMAND with "try to" Probably be good- at least once this eek...Try to get some off	Refers to WILLINGNESS and INDEFINITE FUTURE TIME "I'd like to .. One of these days."
IX	DIRECT COMMAND Re-refers to BENEFIT and TIME as in VIII "Just relax .. try to .. this week even just n hour .. It will make you feel better"	Questions BENEFIT "What if it doesn't?"

FIG. 12.1.   Summary of repeated requests and put offs.

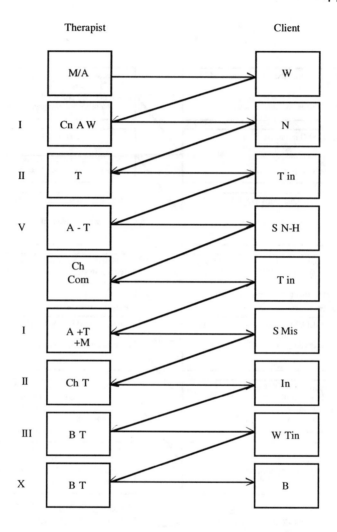

FIG. 12.2. Schematic summary of repeated requests and put off. KEY: *M* = Means; *A*
= Ability; *Cn* = Consequences; *W* = Willingness; *A-T* = Ability Regardless of Time;
Ch = Challenge; *Com* = Competence; *IC* = Indirect Command; *C* = Command; *B* =
Benefit; *N* = need; *Tin* = Indefinite Time; *S N-H* = Spurious Non-Hearing; *S Mis* =
Spurious Misunderstanding; *In* = Indefinite Reference; *R* = Response; *Q* = Question.

DISCUSSION AND IMPLICATIONS

How could this ineffective cycle be broken? How could this stalemate situation become productive? What linguistic skills could enhance the therapist's ability to (a) detect the ruse, (b) diagnose the resistance, and (c) break the cycle? The key lies in recent work by Ainsworth-Vaughn (1998) on negotiating power relationships and on Ervin-Tripp's (1976, 1984) seminal work on American English directives. Ervin-Tripp (1976, 1984) stated that, in exchanges between people of discrepant relative power Need Statements and their subcategory Want Statements rank highest in actual efficacy. According to Ervin-Tripp (1976) a statement such as *I need a match* or *I want you to check the requirements* is more hierarchically effective than imperatives, embedded imperatives, or indirect requests. A focus away from the hearer onto the speaker's needs and wants is predicted by Ervin-Tripp (1976) to secure a direct response of action. We can suggest that if this novice therapist had at any point said to her client, *I want you to tell me how you plan to relax this next week*, or *I need to hear from you what you intend to do in the way of relaxation this week*, then she might have ended the fruitless recursion. She would do well to try a different linguistic means, to vary the form of her requests. We saw this borne out with the successful experienced therapist in example 1b. He said to his client, *I want you to do some homework ... I want you to take 10 relationships*. Recall that the experienced therapist in 1e used a mixture of indirect and direct requests to elicit his client's verbal compliance that she file a report on a rapist in order to face her emotions of outrage, terror, and shame. He was effective in judicious use of linguistic resources. First, however, it is necessary to understand the linguistic means of perpetuating put offs. A study such as this can aid in first identifying the source of communicative non-success, the linguistic means of perpetuating recursion, and a possible linguistic mode for averting or avoiding recursion.

If the insights developed by linguists, sociologists, and anthropologists to account for a recursive sequence of repeated requests and put offs can be shown to apply and are helpful in explaining the communicative non-effectiveness in verbal resistance, then a specialized awareness of the use of language is a valuable tool for psychotherapists to have. The application of sociolinguistic knowledge to a situation such as this can supplement the expertise of a mental health professional and demonstrate how language is used to resist. Perhaps the therapist-in-training here could identify her habits of indirection and mitigating of requests and add to them some other techniques with resulting therapeutic effects.

In conclusion, the concept of resistance, in a specialized context, can be made more precise, more recognizable, and more measurable by using linguistic tools such as discourse analysis. Repeated put offs following indirect requests are one diagnostic device that linguistics offers for gauging resistance. Implications are that specialists need to increase understanding of judicious use of indirection in order to prevent communicative non-effectiveness. The result might be a freer expression of emotion, the hallmark of efficacious therapy.

REFERENCES

Ainsworth-Vaughn, N. (1998) *Claiming power in doctor-patient encounters.* New York: Oxford University Press.

Duranti, A, & Brenneis, D., Eds. (1986). The audience as co-author. *Text, 63,* 3.

Ervin-Tripp, S. (1976). Is Sybil there? The structure of some American English directives. *Language in Society, 5,* 25-66.

Ervin-Tripp, S. (1984, October). *Expanding the notion of indirect requests.* Paper presented at the Ninth Annual Boston University Conference on the Study of Child Language, Boston, MA.

Ferrara, K. (1991). Accommodation in therapy. In H. Giles, N. Coupland, & J. Coupland (Eds.), *Contexts of accommodation: Developments in applied sociolinguistics* (pp. 187-222). Cambridge, UK: Cambridge University Press.

Ferrara, K. (1992). The interactive achievement of a sentence: Joint productions in therapeutic discourse. *Discourse Processes, 15,* 207-228.

Ferrara, K. (1994). *Therapeutic ways with words.* New York: Oxford University Press.

Ferrara, K. (1997). *So*-summary statements: "Speaking for Another" in therapeutic discourse. *Proceedings of the Fourth Annual Meeting of Symposium About Language and Society, Texas Linguistic Forum, 37,* 132-142.

Ferrara, K. (1998). The social construction of language incompetence and social identity in psychotherapy. In D. Kovarsky, J. Duchan, & M. Maxwell, (Eds.), *Disabling clinical and social evaluations: Constructing (in)competence* (pp. 343-361). Mahwah, NJ: Lawrence Erlbaum Associates.

Freud, S. (1960). Psychopathology of everyday life. In J. Strachey (Ed. and Trans.), *The standard edition of the complete psychological works of Sigmund Freud* (Vol. 6, pp. 1-279). London: Hogarth Press. (Original work published 1901)

Gaik, F. (1992). Radio talk-show therapy and the pragmatics of possible worlds. In A. Duranti & C. Goodwin (Eds.), *Rethinking context: Language as an interactive phenomenon* (pp. 271-290). Cambridge: Cambridge University Press.

Gibbs, R. W., Jr. (1979). Contextual effects in understanding indirect requests. *Discourse Processes, 2,* 1-10.

Greenson, R. (1967). *The technique and practice of psychoanalysis* (Vol. 1). New York: International Universities Press.

Grimshaw, A. (1982). Comprehensive discourse analysis: An instance of professional peer interaction. *Language in Society, 11,* 15-47.

Jefferson, G., & Lee, J. R. (1981). The rejection of advice: Managing the problematic convergence of a "troubles-telling" and a "service encounter." *Journal of Pragmatics, 5,* 399-422.

Kolko, D. J., & Milan, M. A. (1983). Reframing and paradoxical instruction to overcome "resistance" in the treatment of delinquent youths: A multiple baseline analysis. *Journal of Consulting and Clinical Psychology, 51*, 655-660.

Labov, W. (1982). Speech actions and reactions in personal narrative. In D. Tannen (Ed.), *Analyzing discourse: Text and talk* (pp. 219-247). Washington, DC: Georgetown University Press.

Labov, W., & Fanshel, D. (1977). *Therapeutic discourse: Psychotherapy as conversation.* New York: Academic Press.

Langs, R. (1981) *Resistance and interventions.* Northvale, NJ: Jason Aronson.

Mahalik, J. R. (1994). Development of the Client Resistance Scale. *Journal of Counseling Psychology, 41*, 58-68.

Restak, R. (1988). *The mind.* New York: Bantam Books.

Schuller, R., Crits-Christoph, P., & Connoly, M. B. (1991). The Resistance Scale: Background and psychometric properties. *Psychoanalytic Psychology, 8*, 195-211.

Weijts, W. (1993). Seeking information. In *Patient participation in gynecological consultings: Studying interactional patterns.* The Hague: CIP Koninklijke Bibbliotheek.

Wodak, R. (1981). How do I put my problem? Problem presentation in therapy and interview. *Text, 1*, 191-213.

APPENDIX A

Background Segments

(To establish overwork and need, as recognizing and commented on by both client and therapist for the client to stop overworking and overworrying.)

A.

T: Last time you talked about that you were working more and more and trying to keep from thinking about it. [Just think]

C: [Yeah] Trying to make myself tired enough that when I lay down I just, you know, went to sleep til the next morning ((clears throat)). And it seemed to work pretty good. Then once I get back into the office environment, u:h=

T: =it doesn't work at all.

C: Yeah, your pressure builds up in there.

B.

T: Are you feeling any differently about the breakup (.) than you were?

C: O::h (4.) I really haven't had a whole lot of time, you know, to just (.5) sit down and think about that. I just-

T: On purpose you haven't?

C: Yeah. Um hmm. One of these days I'm probably, I'm going to *have* to sit down. ((laugh)) Right out at work or something.

C.

C: Like last weekend, I found myself (.3) 3:00 Saturday afternoon in the office working, (.3) and I found myself there Saturday night, and I found myself there Sunday evening and Sunday night.

T: Was it because (.) you wanted to keep busy or because you had to?

C: Uh (.5) a little of both.

T: Yeah.

D.

T: So your overworking is (.3) partly for *you* (.3) in *two* ways (.) so that you'll know what you're doing plus you'll keep yourself busy where you don't think about things and partly because you have to because of the company.

C: Yeah. And uh (.3) when I, when I got out to talk with someone, it's not that (.3) I look at it in a different manner. I'm *not* a salesman (.3) and never have been.

T: Mmhmm.

C: So, but I've just been put in this role and I've done it for six years now, so (.) Seven years now ((clears throat)). But when I go out, I just (.3) I think of myself as a *rep* (.3) between the company and between the customer.

T: Mmhmm.

C: So I get the guff from the company (.3) when they don't pay. And then when something goes wrong there I get the guff from the customer. You know, so::

T: Not a lot of fun (humorous tone))
C: No. A word of advice never did any good.
T: ((weak laugh))

E.

C: The work (.3), the work, it's the type that it's hard to at five-thirty settle down and walk away from. It's because you've got someone out there that's got (.3) you know, a four million dollar company set up that they're running on something *you* sold them, that, you know, and I can't turn my back and walk away.

F.

T: Sounds like you just have all sorts of things thrown at you, all sorts of problems that are headed this way. Really heavy.
C: Yeah. I'll probably collapse ((laugh)) when I walk out the door here.
T: The weight of the [*world* is on you.]
C: [No, it it, yeah.] Yeah. That's really, that's about what I feel like.

Author Index

Subject Index